Building a
LONDON & NORTH WESTERN RAILWAY JUMBO

The Bill Finch Portfolio of Locomotive Details

Mike Williams

Contents

Introduction	2
Ancestry of the Large Jumbos	4
The footplate of the Precedents	9
Tender controls	10
Locomotive details	11
Tender details	42
Index	56
Loco & Tender general arrangement drawings - inside back cover	

The London & North Western Railway Society

Introduction

Bill Finch was born at Aston, Birmingham, in 1906, the son of an electrical engineer. He lived within sight of the Midland Railway line from Birmingham to Derby and his interest in railways developed at an early age. Bill recalls occasional trips to New Street station with his parents and, as a youth, specifically remembers seeing a Midland single, *Princess of Wales*, leaving with a train for Derby.

An interest in engineering led to attendance at metalwork classes at about the age of 12 and, when returning from one such class one afternoon, Bill clearly remembers seeing *Cornwall*, plus directors' saloon, on the LNWR line from Aston to Birmingham. He surmises that C J Bowen Cooke was on board and was returning to Crewe after a shed visit to Aston, but this may be a bit of embroidery!

Bill left school at 14 and was apprenticed as a machinist to Wolseley Motors, then owned by Vickers. His journey to work included a half mile walk adjacent to the LNWR line from Birmingham to Coventry, and it was at this time that his affinity for the North Western was determined. On his morning walk he would see the 7.30am Birmingham to Euston, usually with a Precursor in charge, closely followed by a local to Leamington Spa headed by a Cauliflower, and a local to Coventry, typically headed by a Coal Tank. On his return in the late afternoon, he looked out for a local from Coventry to Birmingham. This appeared to be a 'running-in' turn and brought a variety of locomotives including Experiments and Princes.

Visits to Birmingham New Street were regular at this time, and Bill recalls what big engines the Claughtons appeared to be. But of greater fascination were the Jumbos, which were used for a variety of turns. *Puck*, *Hardwicke* and *Sister Dora* are well remembered, the latter often working locals to Walsall or onto the Harborne Branch.

The 1930s depression cost Bill his job, but he eventually found work as a mould maker with GEC and subsequently as a toolmaker with Joseph Lucas. He lived in Southampton for a short time but returned to his native city before the war. It was at about this time that he became a founder member of the Birmingham Society of Model Engineers (BSME).

Bill's other interests included hill walking, long distance cycling, and gardening (particularly alpines and orchids), but 'live steam' railway modelling became his primary interest. This involved him building a couple of freelance locomotives but, as he progressed, he developed a yearning to construct an accurate model of an LNWR engine. To ensure that sufficient detail could be incorporated, he chose the scale of 1 1/16 in per foot, but he realised that a Claughton in this scale would have been very heavy, so he chose a Jumbo, a small and, in his words, a 'very pretty little engine'.

The frames for the Jumbo were laid in about 1946 and, although Bill spent much of his leisure time in pursuit of the final goal, considerable time was spent in researching and checking detailed information. It was from the late 1940s, and until 1971 when No 1194 *Miranda* finally appeared from the 'paint shop', that most of the drawings presented in this publication were prepared. Perhaps it was not totally coincidental that Bill also retired in 1971, then as tool room foreman of Midland Electrical Manufacturing Co Ltd in Birmingham. The model was exhibited at the 1975 London *Model Engineer* Exhibition, where it won a silver medal.

The information for the sketches in this publication came from a variety of sources. Ernie Dutton, who worked in the Drawing Office at Crewe, lent drawings to Bill in order that he could copy them and make notes. Whilst most of these drawings were eventually claimed by the National Railway Museum, it is thought that some had been lost or destroyed by that time. Where no drawing was available the two preserved LNWR Webb engines *Hardwicke* and the Coal Tank were consulted and details taken from them. Other engines were also examined including *Cornwall* and some of the details herein are for different engines such as Whale Precursors and Webb compounds.

The figure of Bill Finch, on the right, is unmistakable to those who knew him. The other gentleman has not been identified, but the engine is an ex-LNWR Special Tank photographed at Crewe, probably in the late 1950s.
Courtesy Ted Talbot

The name *Miranda* was chosen because Bill had a superb photograph of that engine. The engine was named after the daughter of Prospero in Shakespeare's *The Tempest*. *Miranda* was not made to be just a model in a glass case, and in her early years enjoyed regular outings to the BSME track where she proved to be a good steamer. As time went by, however, her appearances became rarer, as she became too difficult for an ageing man to handle. Not to be outdone, Bill turned his attentions to Gauge 1, and was still turning out locomotives, carriages and wagons in his nineties. The locos, naturally, were live steam models, as Bill could not abide 'electric mice'.

Bill Finch was a founder member of the L&NWR Society in 1973. Sadly, he died in the summer of 1999, shortly before the first edition of this publication could be completed, but his exceptional and superb model of *Miranda* will give great pleasure to future generations and ensure the name 'Bill Finch' will not be forgotten. His son, Les, himself a railway enthusiast, has greatly assisted in preparing this publication and it is comforting to know that the superb model illustrated here is now in safe keeping at the National Railway Museum in York.

Our sincere thanks to Bill and Les Finch for making this publication possible.

1194 *Miranda* standing in Euston station, coaled and watered, ready to back down on to a northbound express. The two whistles can be seen on this picture, as can the lamp socket in the centre of the front buffer beam (which came into use on 1st January 1903) and tender coal rails (introduced 1895).

THE ANCESTRY OF THE 'LARGE JUMBOS'.

Although visually similar to John Ramsbottom's Newton class of 1866, F W Webb's Precedent 2-4-0 of 1874 was in fact a blend of Crewe and Wolverton design practice. The general layout, that is wheelbase, wheel diameter and boiler position, was the same as the Newton but the most successful feature, cylinders with a large 'V' shaped valve chest above, was copied from the Bloomer class 2-2-2s of the Southern Division, itself based on a 2-2-2 design of 1847 by Edward Bury.

A 'V' shaped valve chest allowed the cylinders to be placed closer together leaving room for longer journals in the driving wheel bearings giving longer mileage between works visits. It also gave very direct steam and exhaust passages. The ratio between steam port area and cylinder swept volume was theoretically ideal, and this, together with (for 1874) long travel long lap valves led to a powerful and fast engine.

'Curved Link Motion', as the conventional Stephenson-Howe valve gear used in the Newtons was termed at Crewe, required valves between the cylinders on the same centre line, thus Webb had to design a new valve gear. He could have adapted the curved link gear of the Bloomers but in choosing Alexander Allan's straight link motion, which required less vertical space than the Bloomer gear, he was able to increase the diameter of the boiler without unduly raising its pitch. To allow safe running at speed, the leading wheels were given ¼in side play controlled by inclined slides of the Cartazzi type. Webb actually introduced a 5ft 6in version, the Precursor class, a few months before the first Precedent appeared. A total of seventy Precedents emerged from Crewe works between December 1874 and February 1882.

When the early members of the Newton class fell due for renewal in 1887, an Improved Precedent design was produced. All 96 members of the

Miranda was allocated LMS No. 5068 in 1923, but did not carry the number until February 1928. This picture was probably taken between those dates. The engine has received its third set of buffers, this time the long-tapered Cooke pattern, and the tender has received later pattern axlebox covers. The riveted horizontal joint in the firebox cladding is very noticeable.
Real Photographs W1402

The Improved Precedent No.1194 *Miranda* was built in February 1897, replacing a Precedent of that name which had been built in 1878. In this c1900 view davits can be seen on the tender to hold the old Harrison communication cord. This attached to a second whistle on the cab roof and was discontinued after 1901.

Miranda piloting an unidentified 4-cylinder compound, possibly Jubilee No 1912 *Colossus*, about to depart from Euston.
H Gordon Tidey

A Small Jumbo which Bill Finch remembered seeing was No 2158 *Sister Dora*, photographed in February 1895 in Walsall, the home town of the famous and much-loved nurse in memory of whom the locomotive was named.

*Miranda a*t Chester station in 1921. By this time the engine buffers had received thicker wooden pads and the sand pipe had a wind deflector (introduced in 1914). The second whistle had been removed.
LGRP 22681

Ramsbottom class were 'renewed'. As was standard practice at Crewe, many serviceable parts of the old engines were incorporated in the new ones, in addition to the same number and nameplates suitably over stamped with 'rebuilding' dates. The improvements consisted of: 1in plate frames, in place of 7/8 in; a 150lb per sq in boiler made of ½in steel plates, instead of 140 lb of 11/32 in plate, and fitted with Webb's circular smokebox door in place of Ramsbottom's horizontally hinged version; and 3in thick Siemens-Martin steel tyres in place of the original 2¼in. These differences, combined with improved machining of parts and harder, heavier steel rails, encouraged the enginemen to work the engines harder. The 'Improved Precedents' became famous for their outstanding haulage capacity combined with speed.

In 1893, a year before the last of the Newtons was renewed, the first of the original Precedents was renewed and the rolling programme continued until 1901 when it was terminated still with eight of the original engines not renewed. All these had been rebuilt (reboilered) with all the improvements and, in some cases, new 1in frames as well.

A 'Six foot' version of the 'Improved Precedent' replaced the 90 engines of Ramsbottom's Samson class in 1889-96. Apart from the front of the frames, driving wheel diameter and ashpan and brake rigging, these engines, known henceforth as Small Jumbos, were identical to the larger wheeled ones, known as Large Jumbos.

Two Large Jumbos achieved considerable fame. The first was No 955 *Charles Dickens* which covered one million miles from its completion in February 1882 to September 1891, and eventually clocked up 2,345,107 miles, a total which was never exceeded by steam on the West Coast route. The other famous engine is No 790 *Hardwicke* now preserved in the National Railway Museum at York. One of the regular engines during the 1895 Race to the North, it fell to *Hardwicke* to take over the racing train at Crewe on the final record breaking night of 22nd August. The 141 miles to Carlisle, including the climbs of the Grayrigg and Shap inclines, was covered in 126 minutes at an average speed of 67.2 mph. Between Penrith and Carlisle the average was 74 mph, in order to achieve which, speeds of up to 90 mph must have been reached.

The general design of the 'Straight Link' engines was so sound that the performance of the Jumbos was good even by the standards of 1934, when the last of them was scrapped. Indeed a development of the original Precursor design of 1874, the 5ft 6in Tank 2-4-2T of 1890 of which 160 were built, was so successful that the last example was scrapped in 1955, over a century after the Bury singles and Bloomers appeared.

L. AND N. W. R. COMPANY'S ENGINE "CHARLES DICKENS," NO. 955.

SIR,—In 1891 you kindly published a letter after the above engine had run one million of miles. On the 5th instant she completed two millions of miles, and you may think it of sufficient public interest to publish the further letter I herewith enclose.

London and North-Western Railway,   F. W. WEBB.
Locomotive Department, Crewe, August 22nd.

Sir,—In September, 1891, my Manchester friends were informed that on the 12th of that month I completed my millionth mile of travel, that all my energies were unimpaired, that I felt no symptoms of decay, and that I was about to take a week's holiday. That autumn holiday proved immensely beneficial: I was soon in evidence again, and with other subsequent brief periods off duty for recuperation and rest I have been able to keep the road almost uninterruptedly, until I am now in the proud position to announce that on the 5th instant, before finishing my 5312th trip from Manchester to London and back, and when about three-eighths of a mile on the north side of Bramhall Station, on the Macclesfield line, I completed, with 186 other trips, my two millionth mile of active service for the London and North-Western Railway Company and their supporters. It is not, I hope, too egotistical to say that this is a good record, which has not, so far as I know, been excelled by any British sister, or by any American or European engine yet constructed. I was turned out of my owners' works at Crewe on February 6th, 1882, sent to Longsight to run as often as I could between Manchester and London as the minimum work for a day, with David Pennington and Leigh Bowden as my guides, taking until July, 1888, the 7.45 a.m., then altered to the 8.15 a.m., and in July, 1889, to the 8.30 a.m. train out of Manchester and returning with the 4 o'clock out of London; and right faithfully did my guides conduct and control me. The many thousands of passengers who confidently seated themselves in the coaches I drew were all, I am glad to say, safely carried to their destinations. After a few years, David's eyesight was injured, and on March 17th, 1886, I lost my faithful guide, but Josiah Mills, who succeeded him, together with Leigh Bowden, who has been with me throughout my career, have been unremitting in their attentions, Leigh guiding me over quite half the two million miles I have travelled. He is still in excellent health, and is capable, like my designer, of doing some more important work. Two millions of miles in 20 years and 181 days! How did I effect it? Well, during the time I drank 204,771 tons of water, and to develop my energies and make them equal to the duty required of me, I consumed 27,486 tons of coal!; but although in the interval my travelling pace was increased by degrees from 42 to 50¼ miles per hour, and the weight I had to haul was appreciably increased from July, 1898, to enable breakfast to be served on board, and from July, 1899, to admit of tea being served on the return journey, my consumption of fuel, including the raising of steam each day, did not exceed 32 lb. per mile. Of course, to accomplish this my digestive organs were kept in a healthy condition. The advantages of the standardisation, which is a feature of the Crewe works, and which admits of the speedy interchangeability of parts, reduced enforced idleness to about 12 per cent. of working time. The friends I had always carried safely had even by September 7th, 1886, when my performance had just exceeded half a million of miles, christened the trains I worked the "Charles Dickens," and had ceased mentioning the times of departure when making their travelling arrangements. The cost of maintaining me in efficiency only averaged 1·28d. per mile run. I venture to hope the public, especially of Manchester and those engaged in the mechanical engineering profession, will be interested in my history. I feel now that at my time of life, although still in excellent condition, with all the exacting provisions of the age—breakfast, luncheon, tea, and dining saloons, corridor trains, reservoirs for gas for cooking, steam heating, electric lighting, and generally more luxurious accommodation—I ought to give way to a more powerful comrade, and ask my owners for the indulgence of lighter duty. I thank all the travelling public who have entrusted me with their safety and for the confidence they placed in me; and I hope my successor will also obtain and deserve their support, and win for herself a name for punctuality and general reliability which I trust I have succeeded in doing for myself during the past of my life.—I have the honour to be, Sir, your obedient servant,   "Charles Dickens," L. and N. W. R.,
Engine No. 955.

Engine No. 955.

Longsight Station, August 21st.   1902   (From the 'Engineer')

The cab of *Hardwicke*, photographed in the 1960s in the British Transport Museum in Clapham This is included for comparison with Bill Finch's model version overleaf. Peter Chatham

No, this is not *Hardwicke* again, but the controls of Bill Finch's 5-inch gauge *Miranda*. Simon Fountain

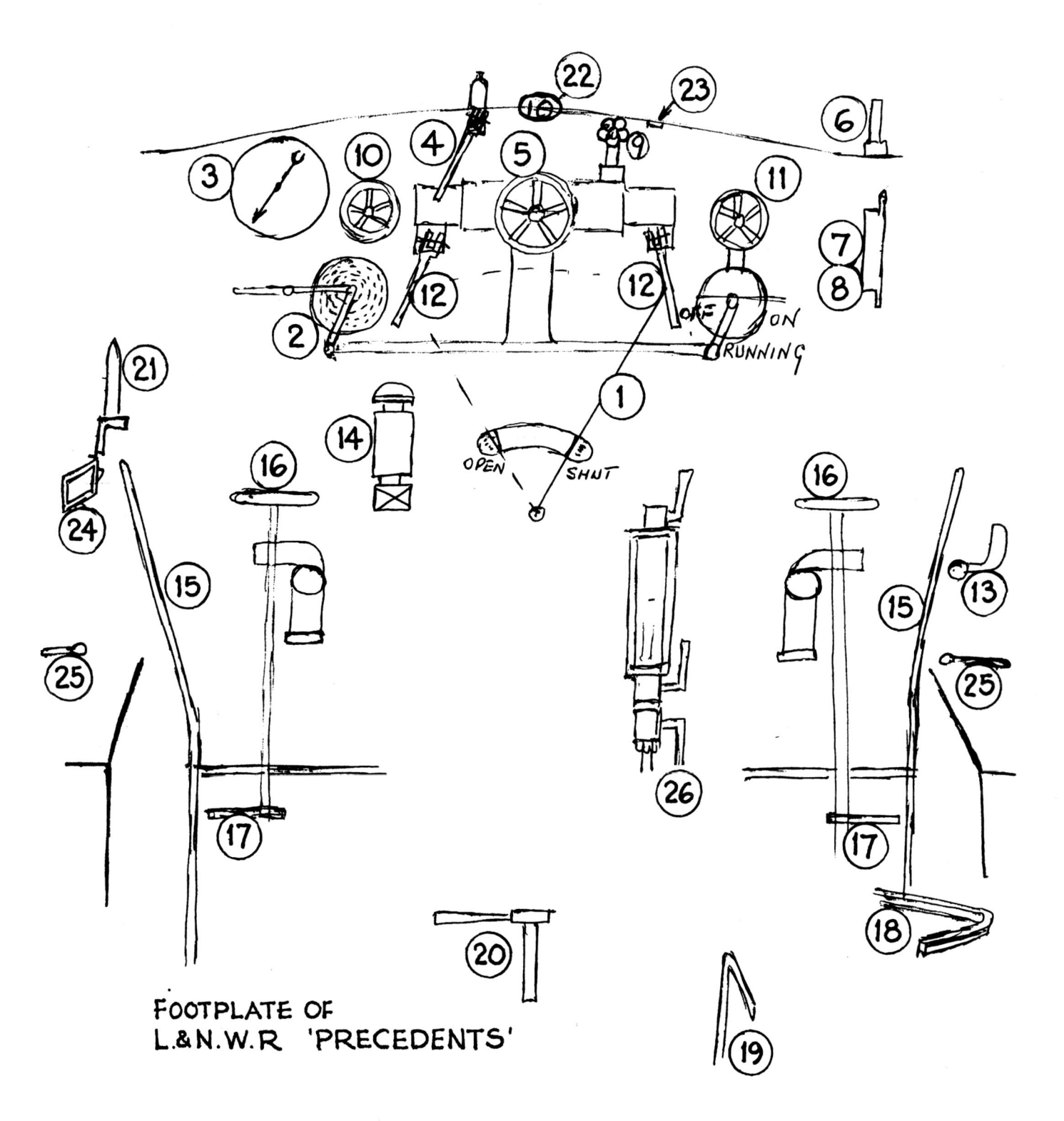

FOOTPLATE OF L.&N.W.R 'PRECEDENTS'

1	Regulator	14	Steam Chest Lubricator (via hollow stay)
2	Vacuum Brake	15	Leading Driving Wheel Sanders
3	Vacuum Gauge	16	Reverse Steam back through Injectors
4	Whistle Valve (pull)	17	Injector Water Control
5	Carriage Warming Valve	18	Drain Cocks
6	Carriage Warming Relief Valve	19	Dampers
7	Carriage Warming Pressure Gauge (front)	20	Boiler Blow Down
8	Carriage Warming Pressure Gauge (rear)	21	Staff Carrier
9	Boiler Pressure Gauge Stop Cock	22	Shed Plate
10	Steam to lubricator	23	Holder for Boiler Washout Reports
11	Steam to Brake Ejector	24	Alternative Holder for above
12	Live Steam Injectors (pull)	25	Rear Sanders (in cab)
13	Blower Valve	26	Water Gauge Blowdown

TENDER CONTROLS

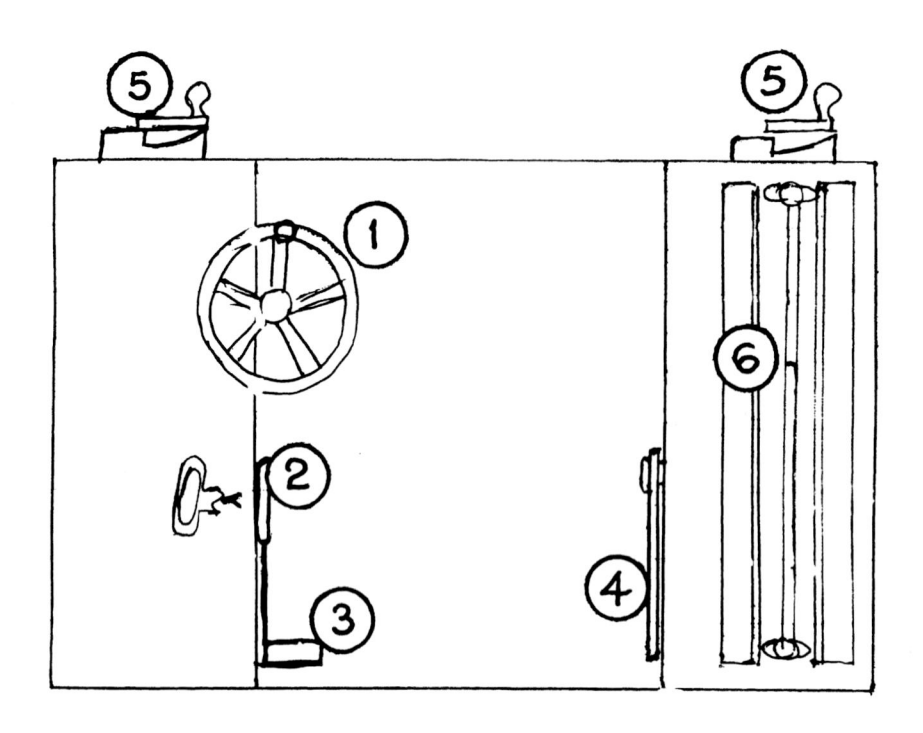

1 Hand Brake	4 Flap Plate Catch
2 Water Scoop Handle	5 Water Controls to Injectors
3 Foot Rest (Water scoop lift aid)	6 Water Level Indicator

Unfortunately we have no photograph which clearly shows the tender controls of Bill Finch's model. These views are of the tender of *Hardwicke* in the British Transport Museum, Clapham in the 1960s. (See Pages 47 and 48)

Peter Chatham

FOOTPLATE

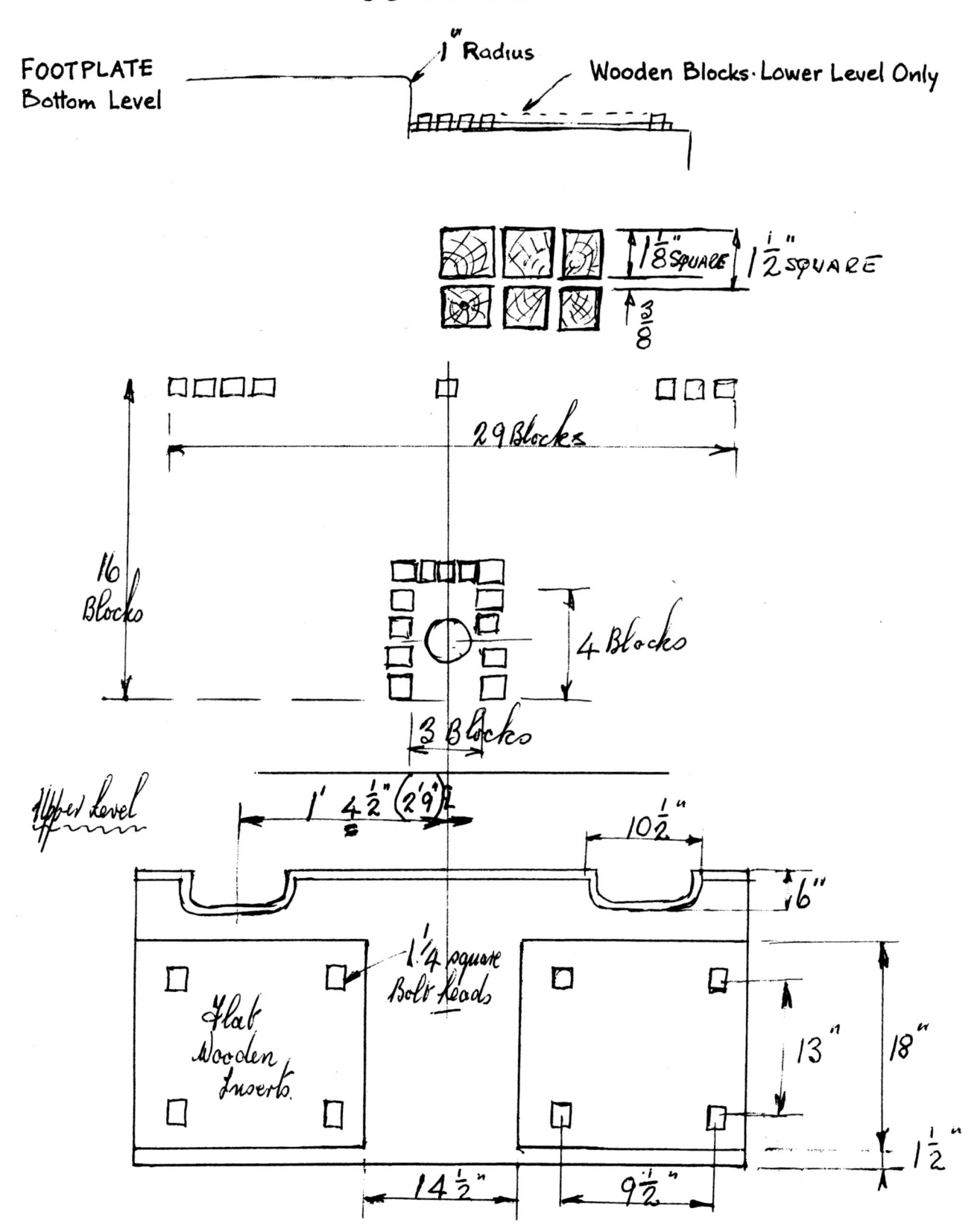

FOOTPLATE Bottom Level

1" Radius

Wooden Blocks·Lower Level Only

$1\tfrac{1}{8}"$ SQUARE $1\tfrac{1}{2}"$ SQUARE

$\tfrac{3}{8}$

29 Blocks

16 Blocks

4 Blocks

3 Blocks

Upper Level

1' 4½" (2'9")

10½"

6"

1¼ square Bolt Heads

Flat Wooden Inserts

13"

18"

1½"

14½"

9½"

11

SHED PLATE

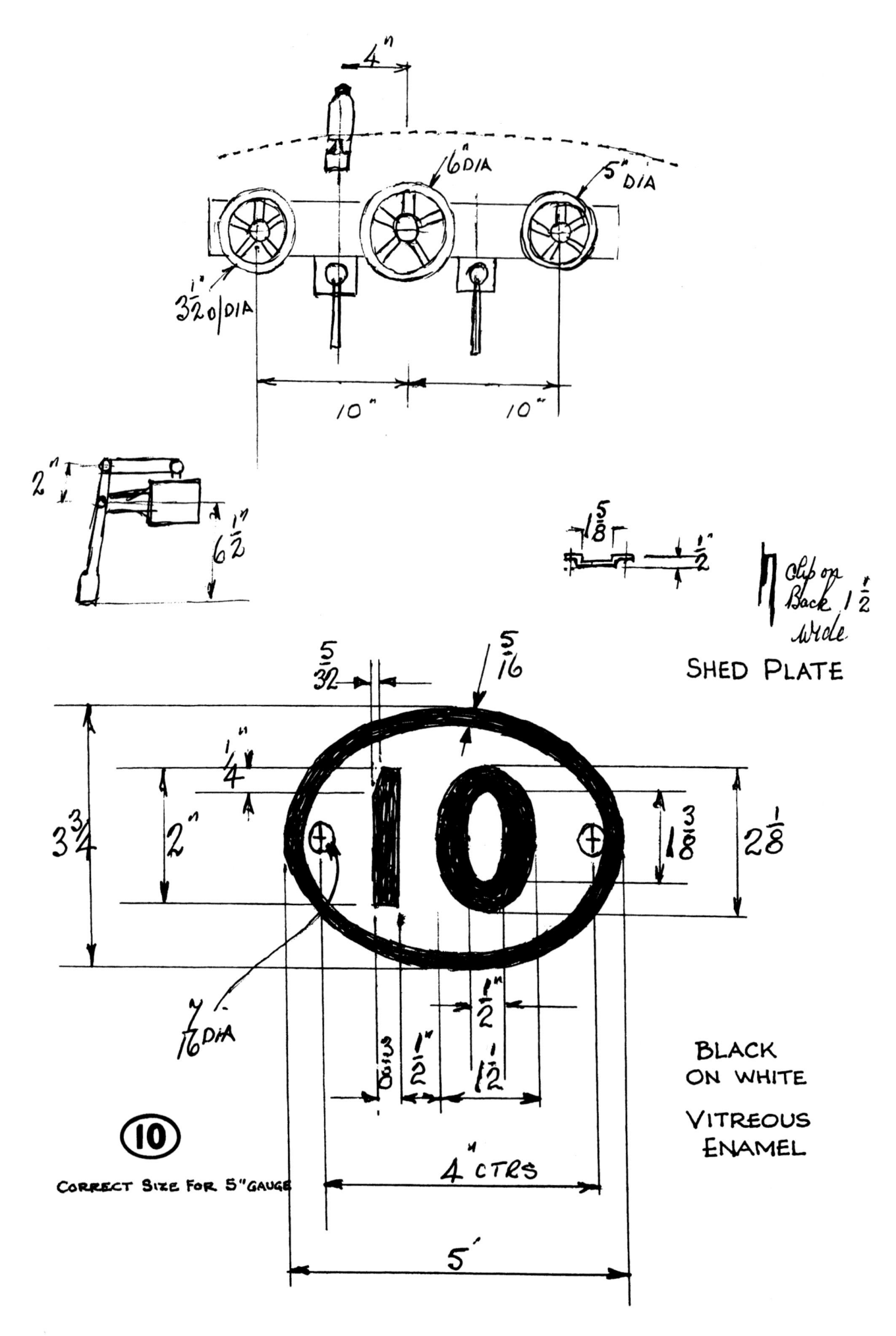

BLACK ON WHITE

VITREOUS ENAMEL

⑩ CORRECT SIZE FOR 5" GAUGE

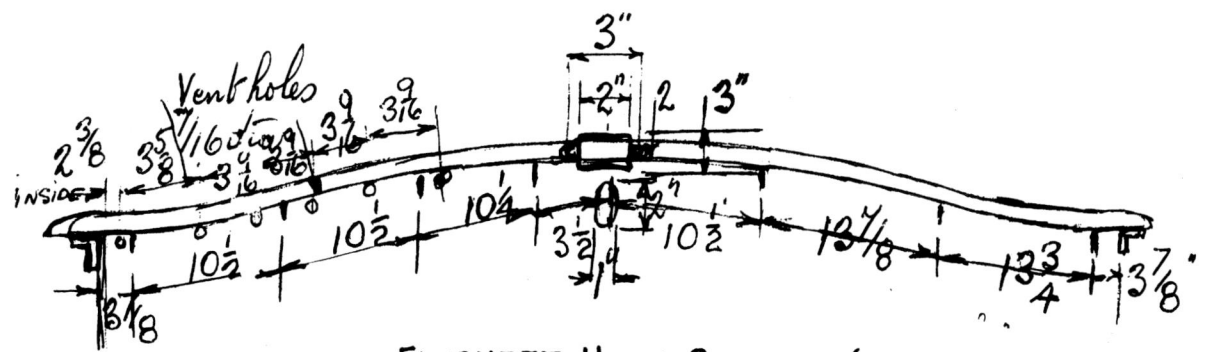

FLYSHEET HOOK CENTRES (approx.) FROM HARDWICKE

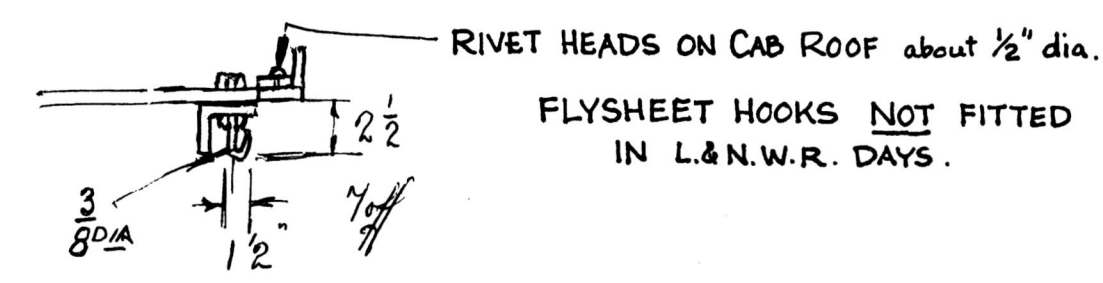

RIVET HEADS ON CAB ROOF about ½" dia.

FLYSHEET HOOKS <u>NOT</u> FITTED IN L.&N.W.R. DAYS.

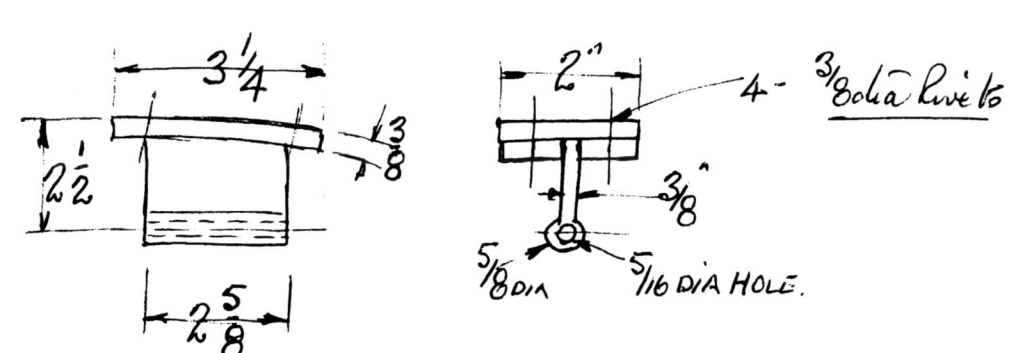

WINDOW CATCHES UNDER CAB ROOF (1 LH 1 RH)

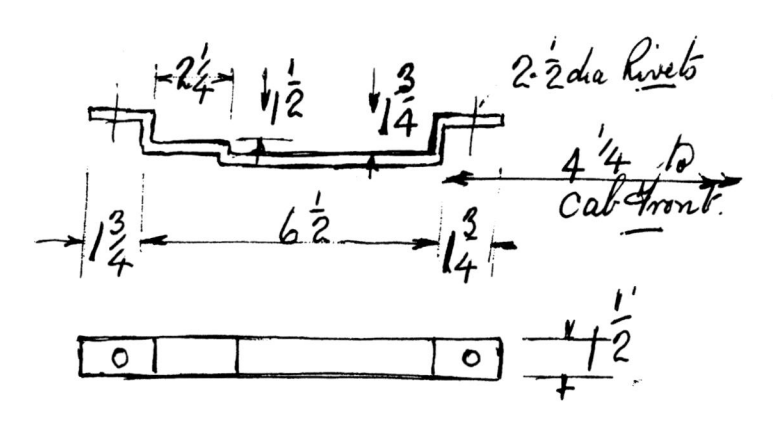

HOLDER FOR BOILER REPORTS

One only, under right side of cab roof in line with window catch on 4ft. centres

BLOWDOWN COCK OPERATING HANDLE

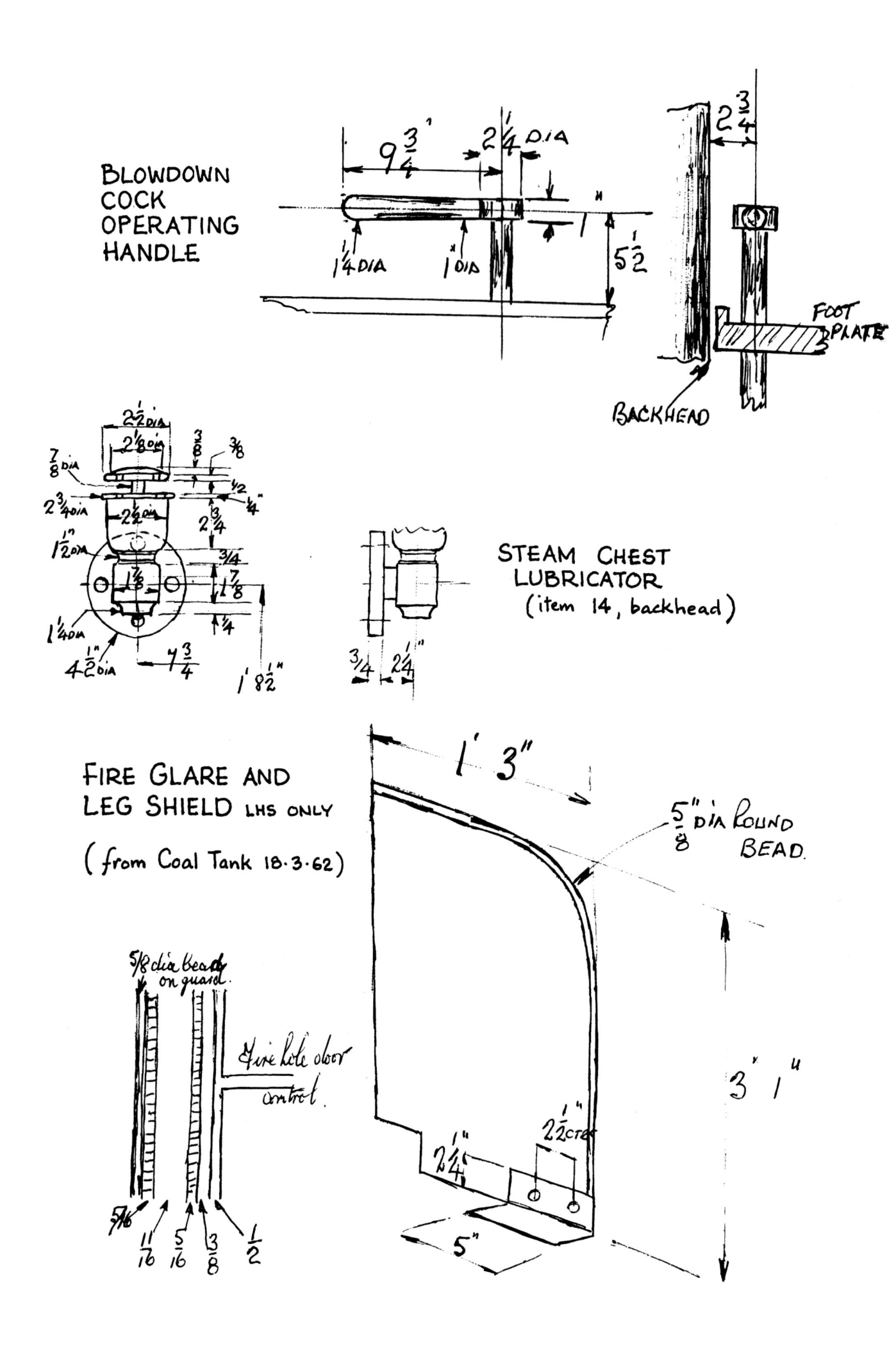

STEAM CHEST LUBRICATOR (item 14, backhead)

FIRE GLARE AND LEG SHIELD LHS ONLY

(from Coal Tank 18.3.62)

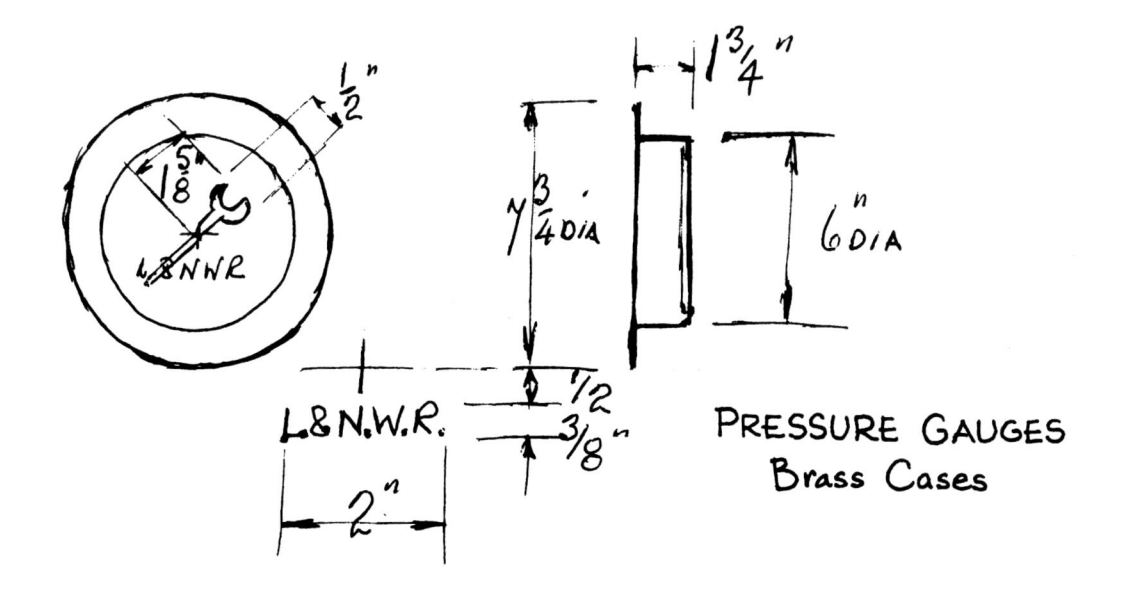

PRESSURE GAUGES
Brass Cases

BLOWER HANDLE on 'Hardwicke'

Full cycle possible without Vacuum Pipe.

Blower Handle on HARDWICKE.

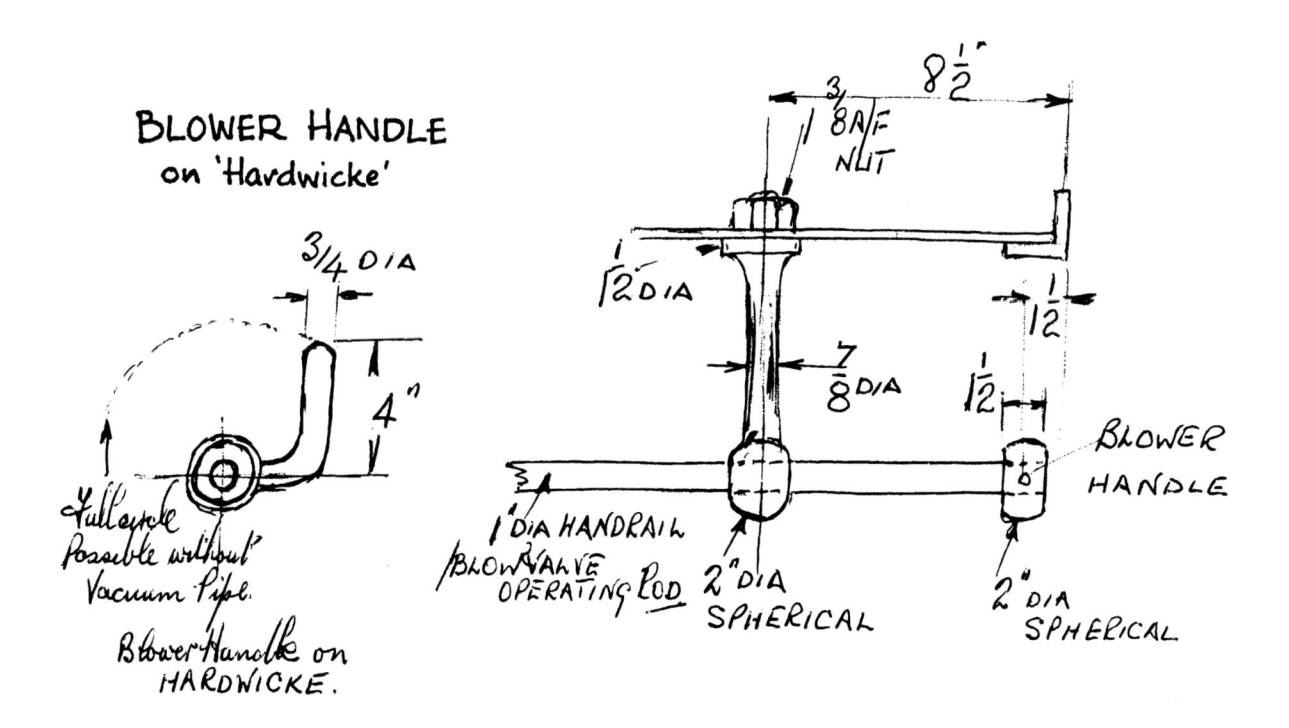

BLOWER HANDLE on the 'Coal Tank' 18.3.62

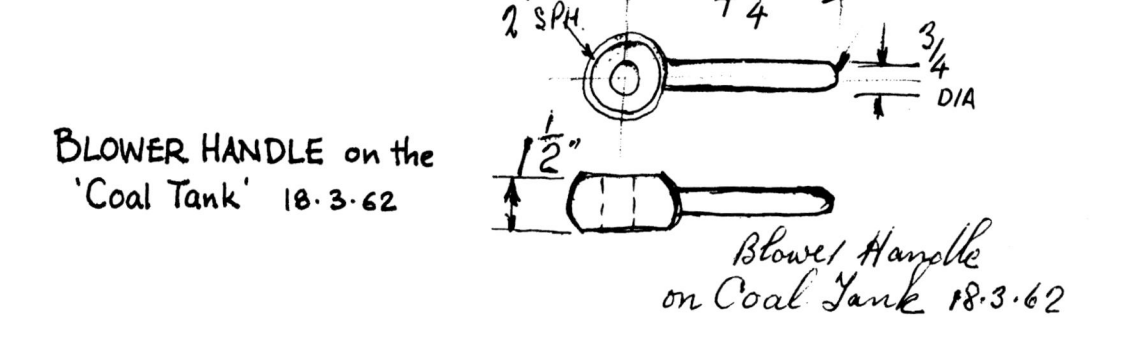

Blower Handle on Coal Tank 18.3.62

VACUUM BRAKE VALVE

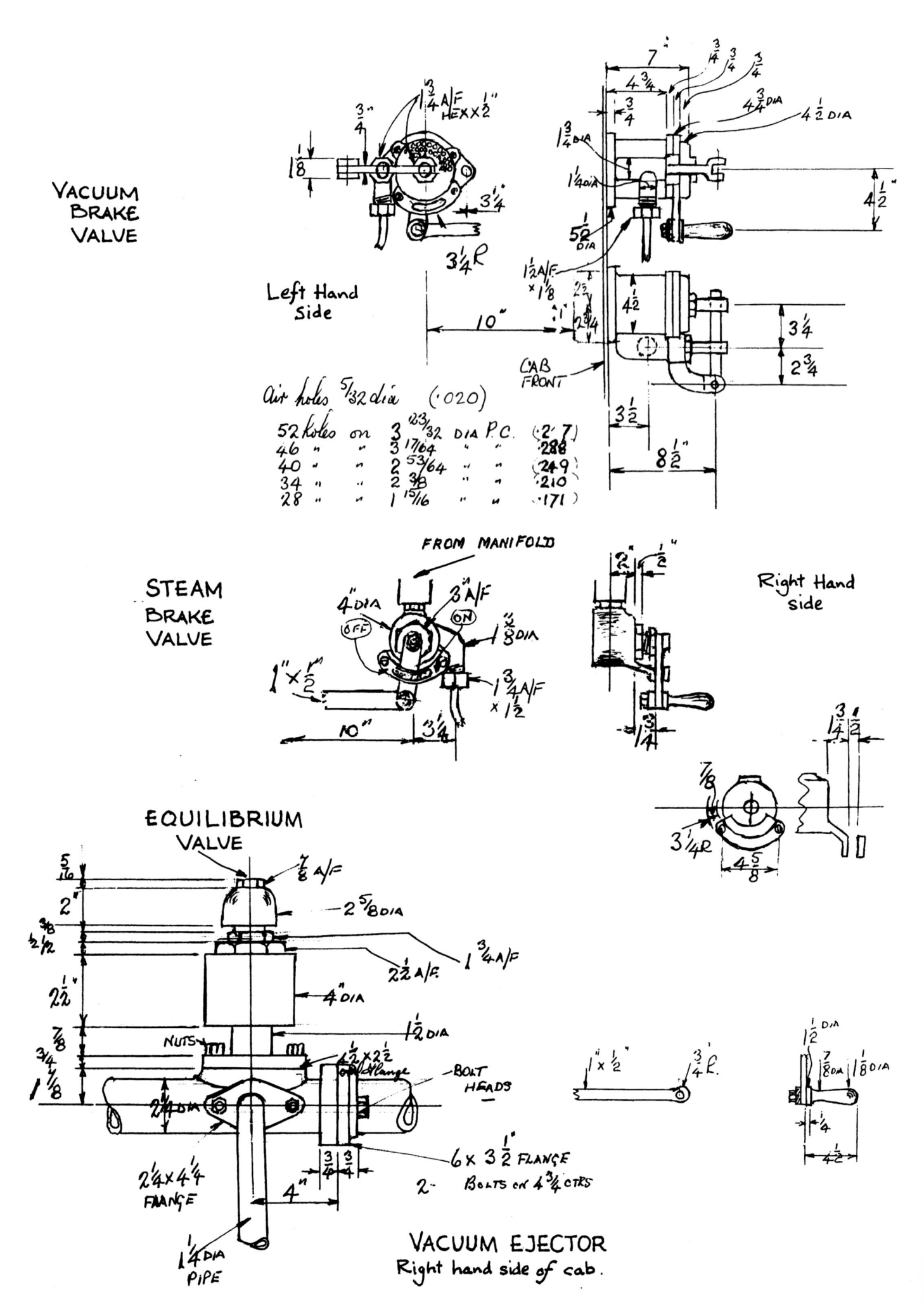

Left Hand Side

Air holes 5/32 dia (·020)
52 holes on 3 23/32 dia P.C. (2·7)
46 "   " 3 17/64 "   " ·288
40 "   " 2 53/64 "   " ·249
34 "   " 2 3/8   "   " ·210
28 "   " 1 15/16 "   " ·171

STEAM BRAKE VALVE

FROM MANIFOLD

Right Hand side

EQUILIBRIUM VALVE

VACUUM EJECTOR
Right hand side of cab.

FIRE HOLE DOOR

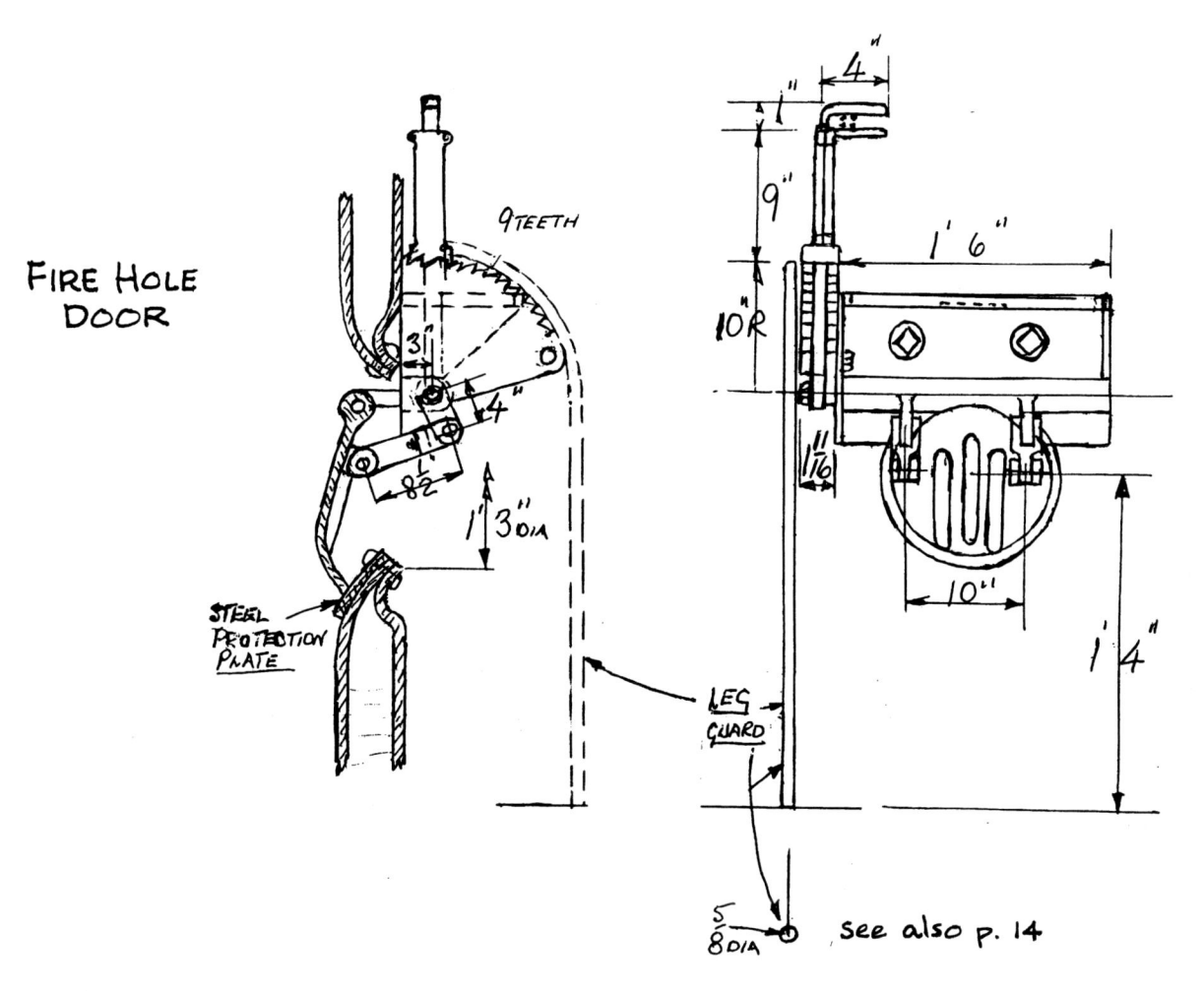

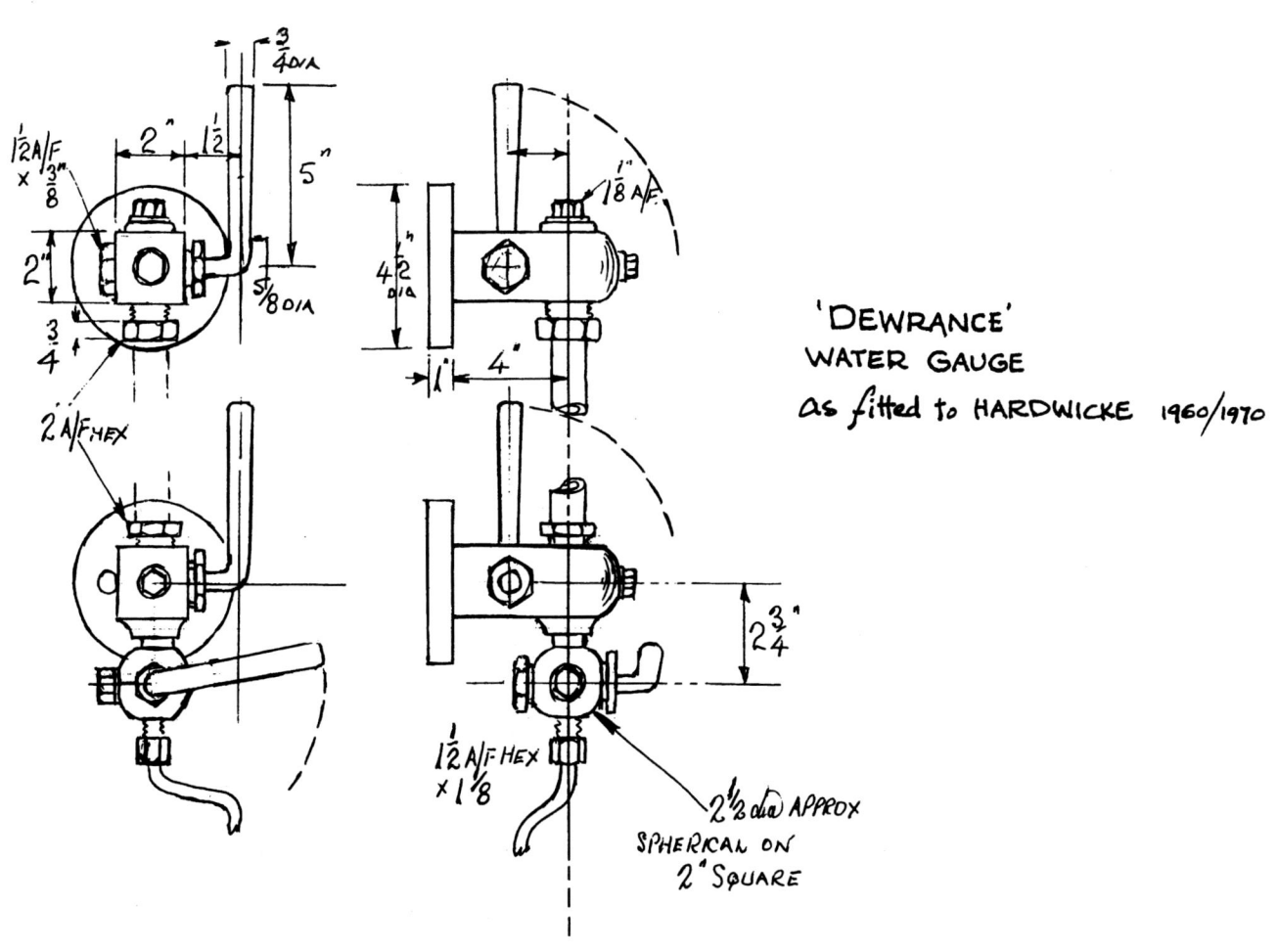

'DEWRANCE' WATER GAUGE
as fitted to HARDWICKE 1960/1970

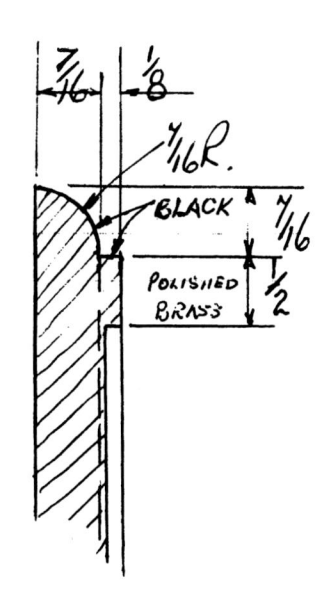

NUMBER PLATES

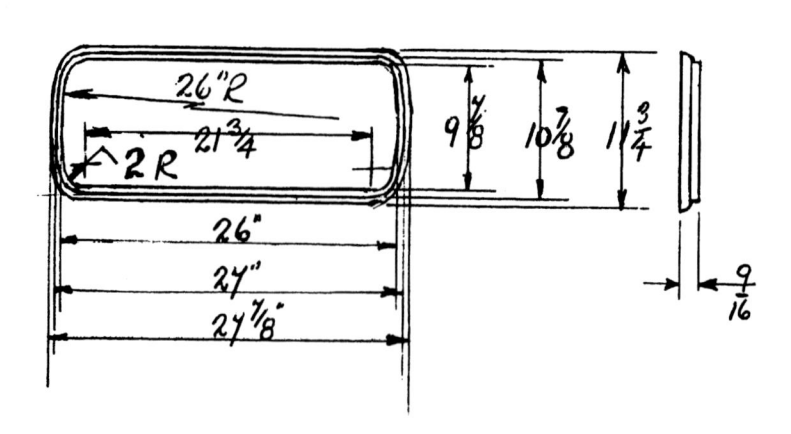

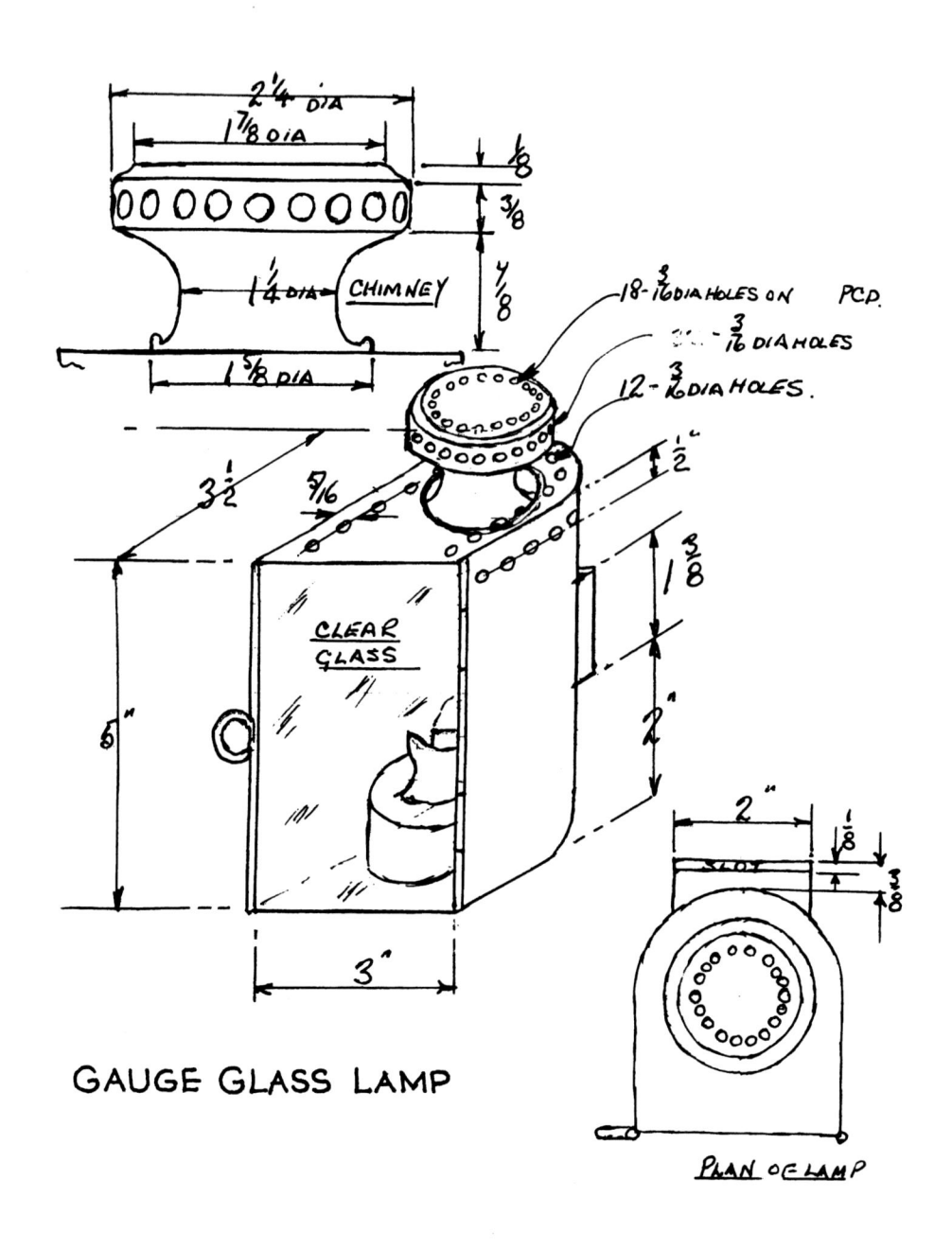

GAUGE GLASS LAMP

PLAN OF LAMP

CAB SPECTACLES

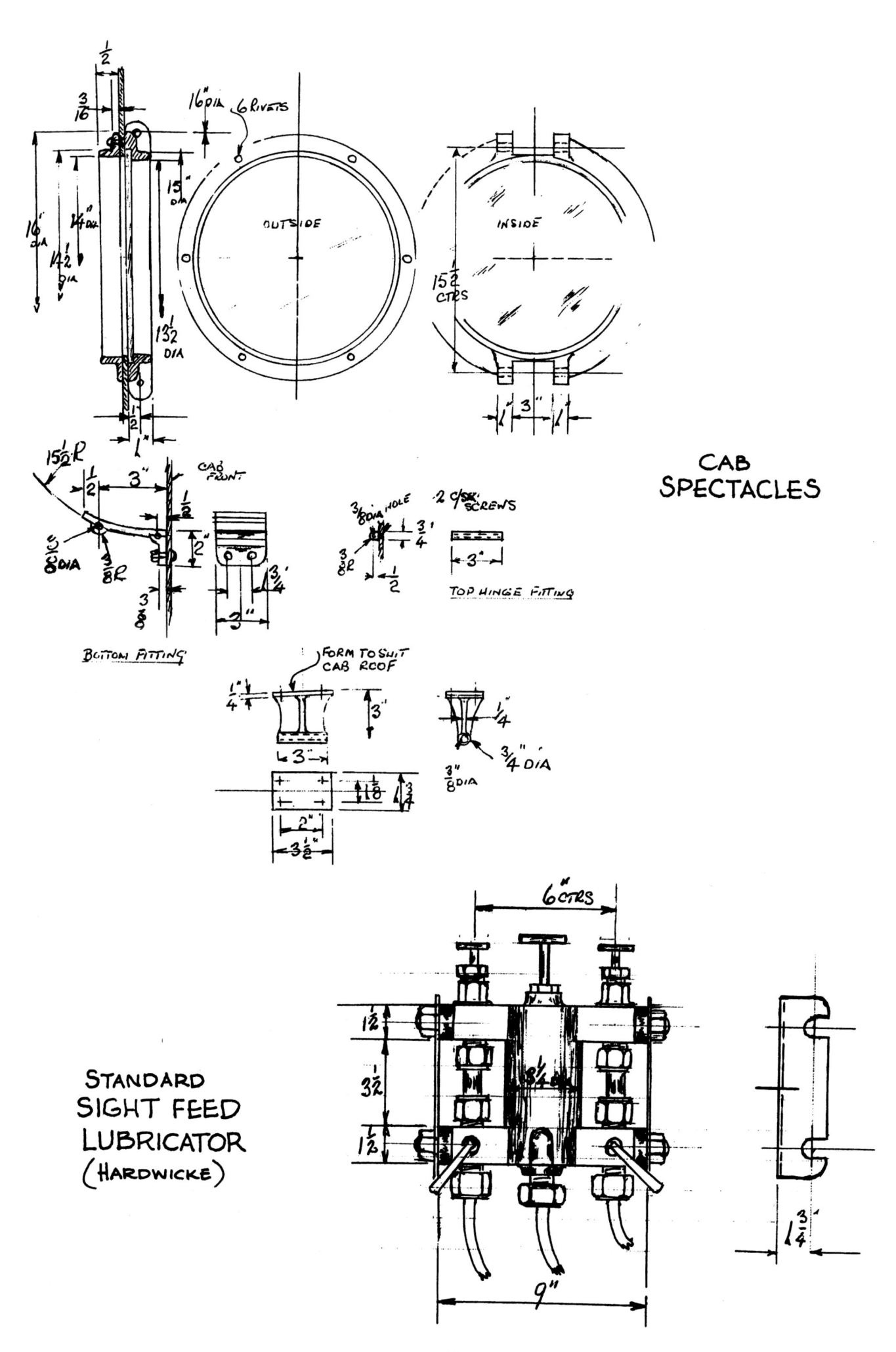

STANDARD SIGHT FEED LUBRICATOR (HARDWICKE)

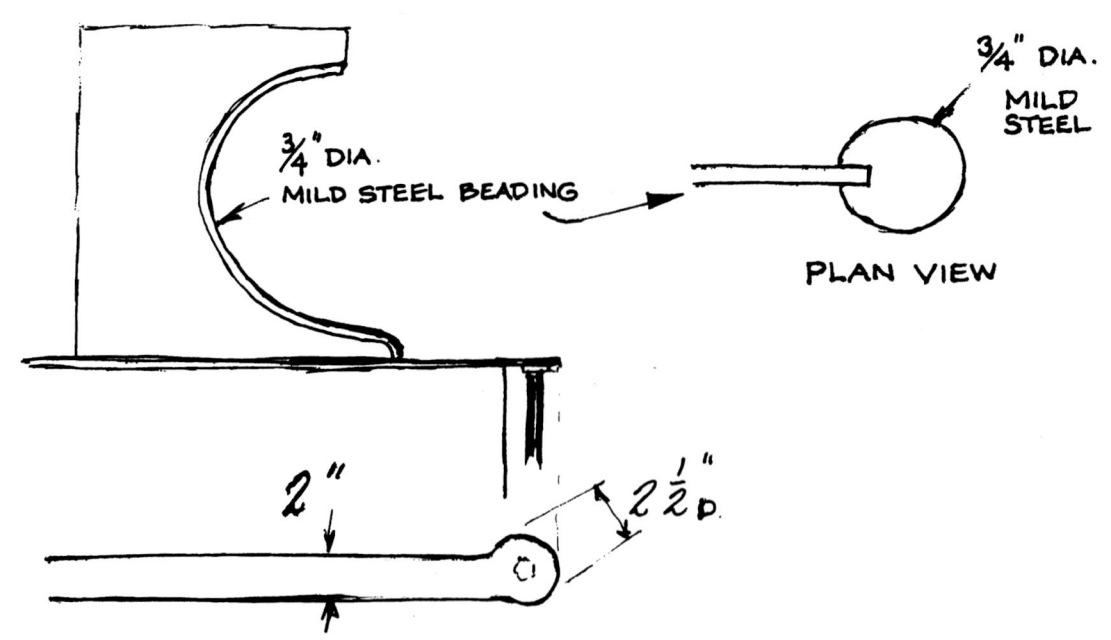

¾" DIA. MILD STEEL BEADING

¾" DIA. MILD STEEL

PLAN VIEW

2"

2½" D.

CAB BEADING AND HANDRAILS

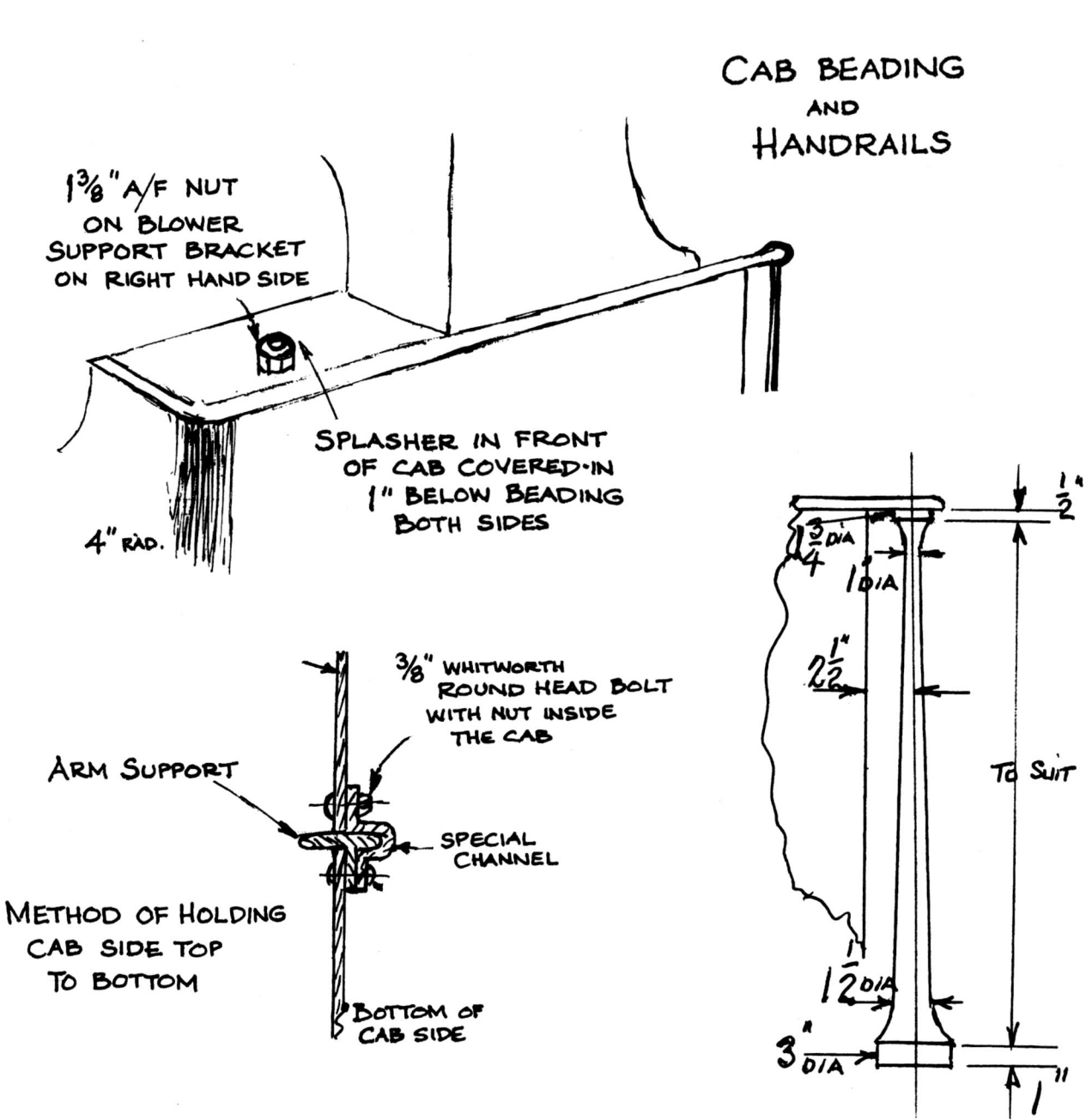

1⅜" A/F NUT ON BLOWER SUPPORT BRACKET ON RIGHT HAND SIDE

SPLASHER IN FRONT OF CAB COVERED·IN 1" BELOW BEADING BOTH SIDES

4" RAD.

⅜" WHITWORTH ROUND HEAD BOLT WITH NUT INSIDE THE CAB

ARM SUPPORT

SPECIAL CHANNEL

METHOD OF HOLDING CAB SIDE TOP TO BOTTOM

BOTTOM OF CAB SIDE

½"
¾" DIA
1" DIA
2½"
TO SUIT
1"
1½" DIA
3" DIA
1"

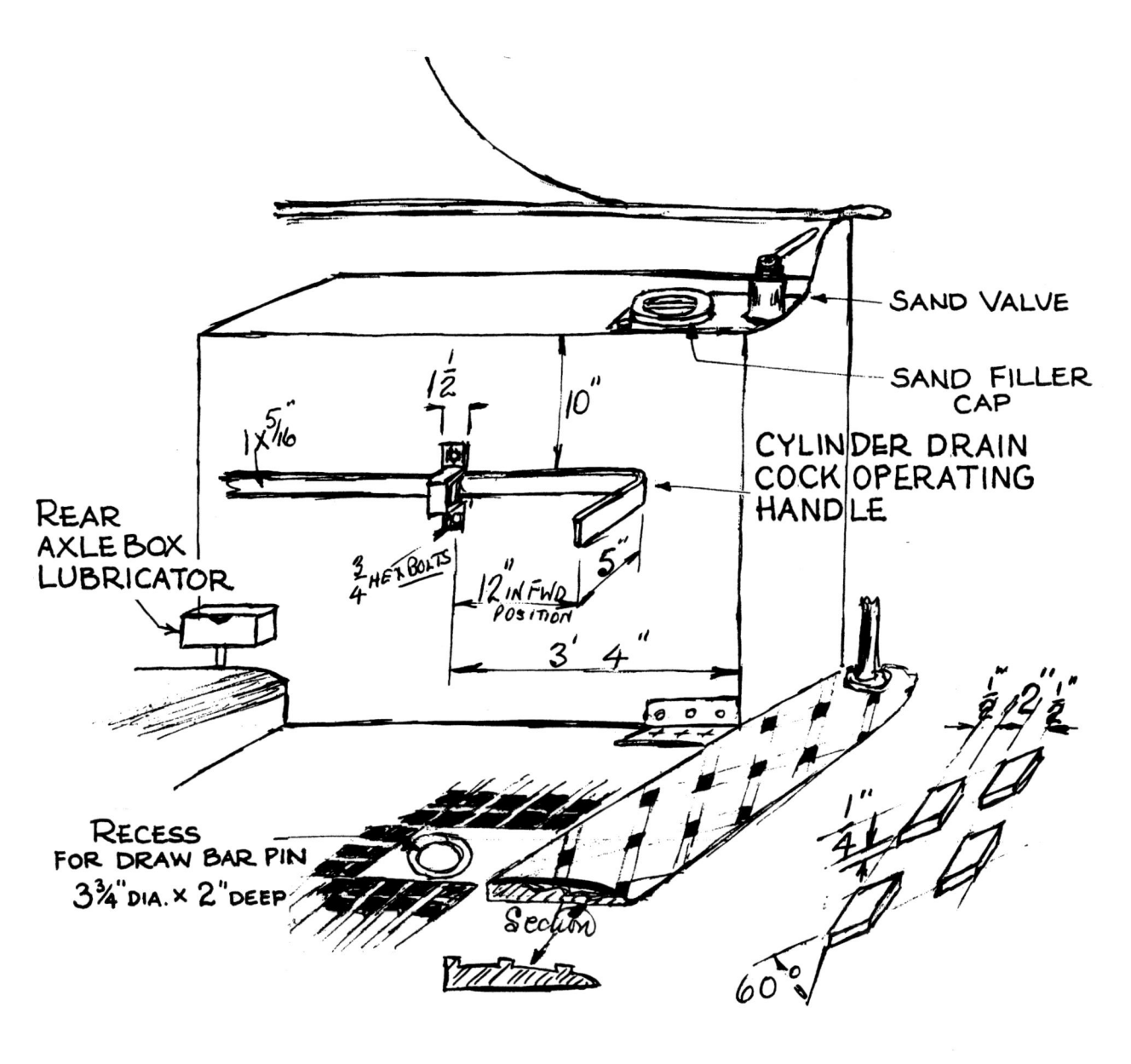

- SAND VALVE
- SAND FILLER CAP
- CYLINDER DRAIN COCK OPERATING HANDLE

1½"
10"
1 x 5/16"
¾" HEX BOLTS
12" IN FWD POSITION
5"
3' 4"

REAR AXLE BOX LUBRICATOR

RECESS FOR DRAW BAR PIN 3¾" DIA. X 2" DEEP

Section

½" 2" ½"
1" ¼"
60°

REAR CORNER OF CAB

4" RAD.
9"
1" A/F SQUARE BOLT

2½" x 5/8"
5/8" Bolts
SAND PIPE 2¼" DIA
ENGINE STEPS RHS.

4/8" DIA
3½" 4½" 5"
5/8" ROUND HEAD BOLT

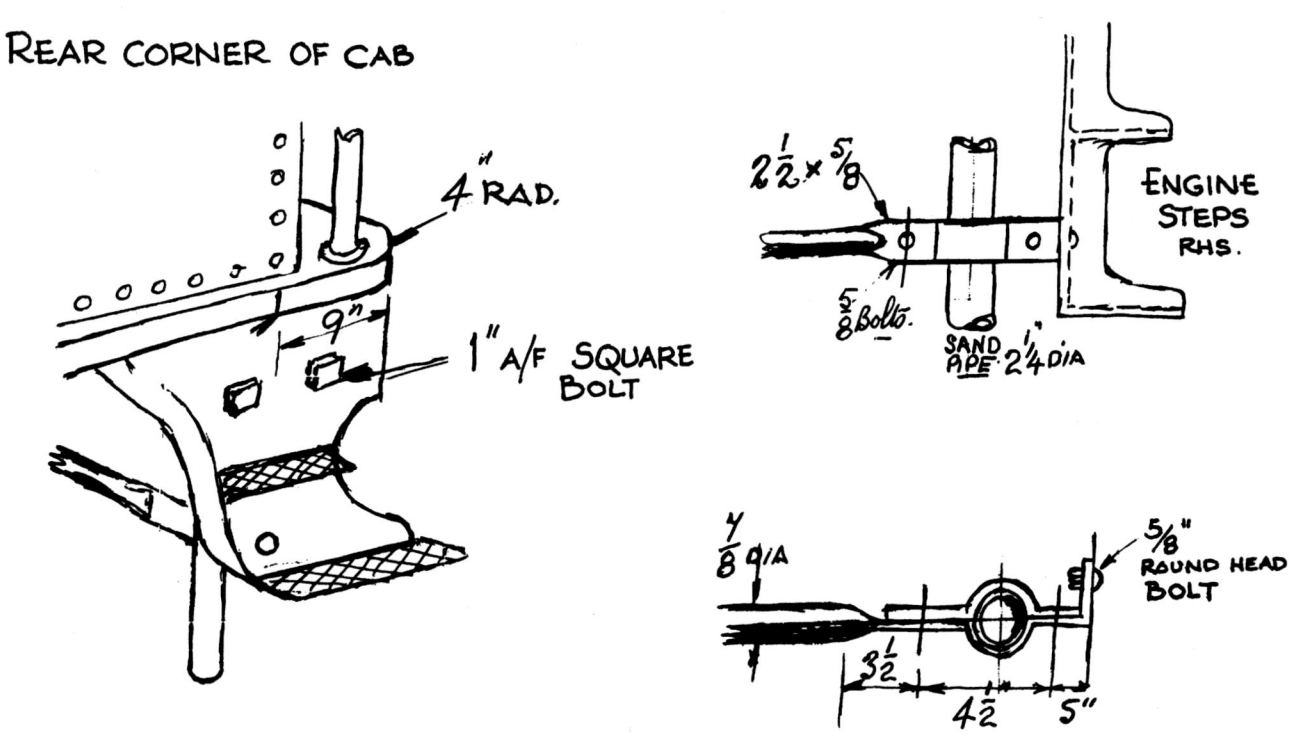

WHISTLE

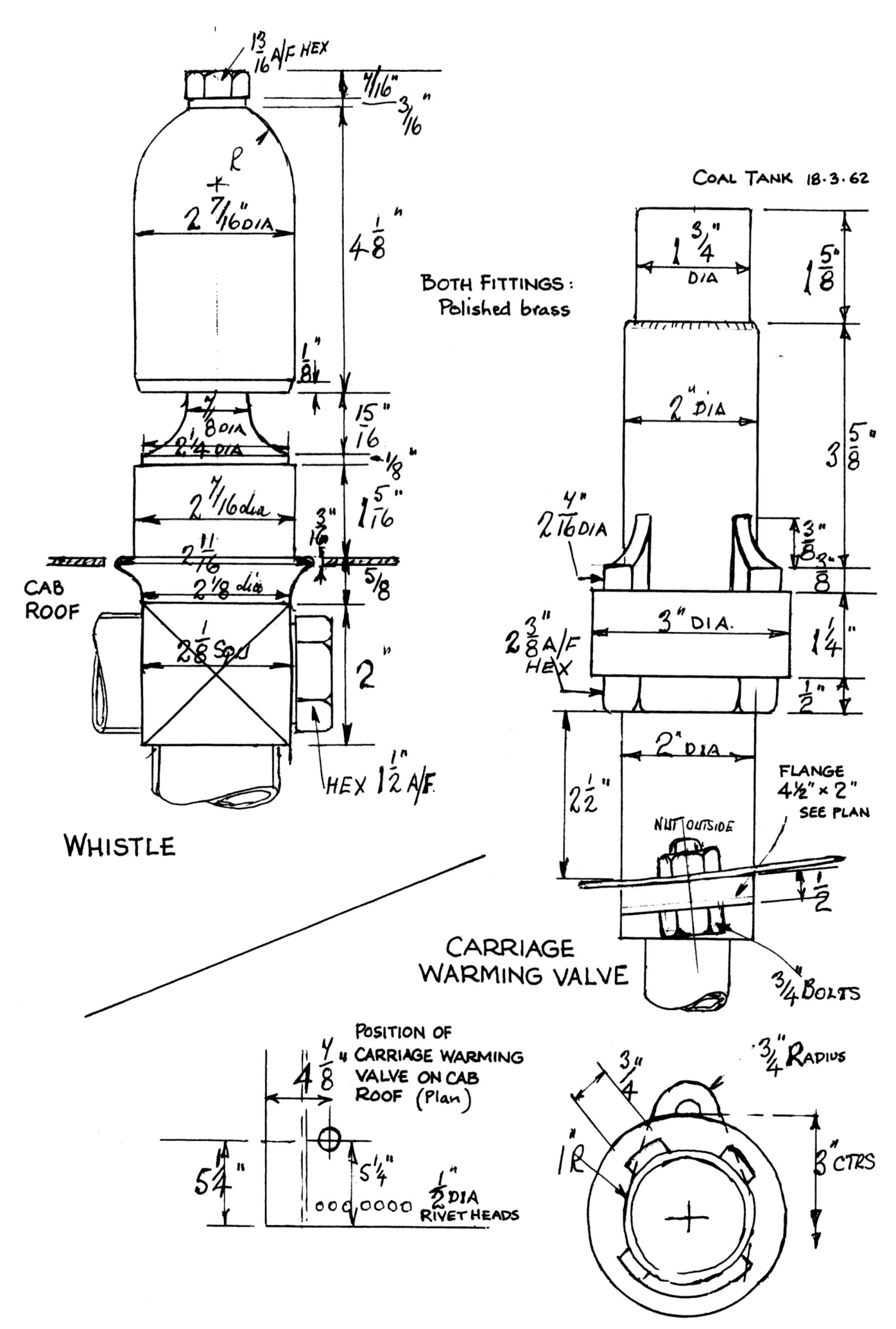

- 13/16 A/F HEX
- 7/16"
- 3/16"
- R
- 2 7/16" DIA
- 4 1/8"
- 1/8"
- 7/8 DIA
- 2 1/4 DIA
- 15/16"
- 1/8"
- 2 7/16 dia
- 3/16"
- 1 5/16"
- 2 11/16
- 2 1/8 dia
- 5/8"
- CAB ROOF
- 2 1/8 Sq
- 2"
- HEX 1 1/2 A/F.

BOTH FITTINGS: Polished brass

CARRIAGE WARMING VALVE

COAL TANK 18.3.62

- 1 3/4" DIA
- 1 5/8"
- 2" DIA
- 3 5/8"
- 2 4/16 DIA
- 3/8"
- 3/8"
- 2 3/8" A/F HEX
- 3" DIA.
- 1 1/4"
- 1/2"
- 2" DIA
- 2 1/2"
- NUT OUTSIDE
- FLANGE 4 1/2" × 2" SEE PLAN
- 1/2"
- 3/4 BOLTS

POSITION OF CARRIAGE WARMING VALVE ON CAB ROOF (Plan)

- 4 4/8"
- 1/4"
- 5 1/4"
- 5 1/4"
- 1/2" DIA RIVET HEADS

- 3/4"
- 3/4" Radius
- 1" R
- 3" CTRS

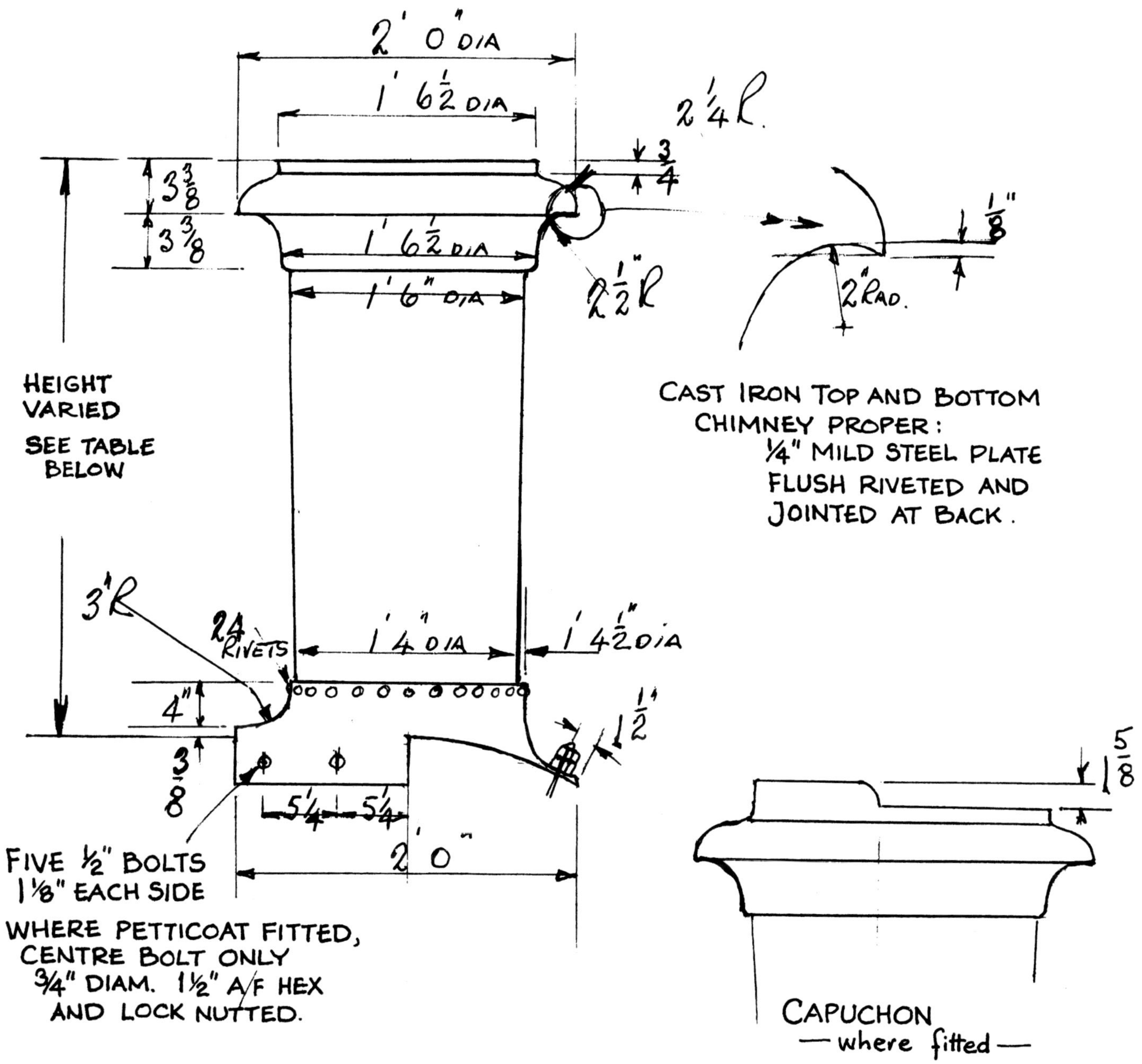

2' 0" DIA
1' 6½ DIA
2¼ R.
¾
3⅜
3⅜
1' 6½ DIA
1' 6" DIA
2½ R
1/8"
2" RAD.
HEIGHT VARIED
SEE TABLE BELOW
CAST IRON TOP AND BOTTOM
CHIMNEY PROPER:
¼" MILD STEEL PLATE
FLUSH RIVETED AND
JOINTED AT BACK.
3' R
24 RIVETS
1' 4" DIA
1' 4½" DIA
4"
1½"
3/8
5¼ 5¼
2' 0"
FIVE ½" BOLTS
1⅛" EACH SIDE
WHERE PETTICOAT FITTED,
CENTRE BOLT ONLY
¾" DIAM. 1½" A/F HEX
AND LOCK NUTTED.
1 5/8
CAPUCHON
— where fitted —

3'-9" SPECIAL DX
COAL ENGINES
0-6-2 COAL TANK
5'-6" 2-4-2 TANK
2-2-2 PROBLEM

3'-3" 6'-0" & 6'-6" JUMBOs
18" CAULIFLOWERS & 0-6-2T
2-2-2-0 EXPERIMENT
2-2-2-0 DREADNOUGHT

2'-10" JEANIE DEANS
GREATER BRITAIN & J. HICKS
JUBILEE 4cyl. COMPOUND
A, B, C. C1 0-8-0

2'-5" ALFRED THE GREAT
4-6-0 4cyl. COMP. (Bill Baileys)

2'-0" 4-4-0 PRECURSORS
4-4-0 GEORGE THE FIFTHS
0-8-0 D, G, G1, G2
0-8-0 & 0-8-2 TANKS
0-8-4 TANKS
4-6-2 TANKS

1'-10" 4-4-2 PRECURSOR TANK
4-6-0 EXPERIMENT
4-6-0 PRINCE of WALES
4-6-0 19" GOODS

1'-8" 4-6-0 CLAUGHTONs

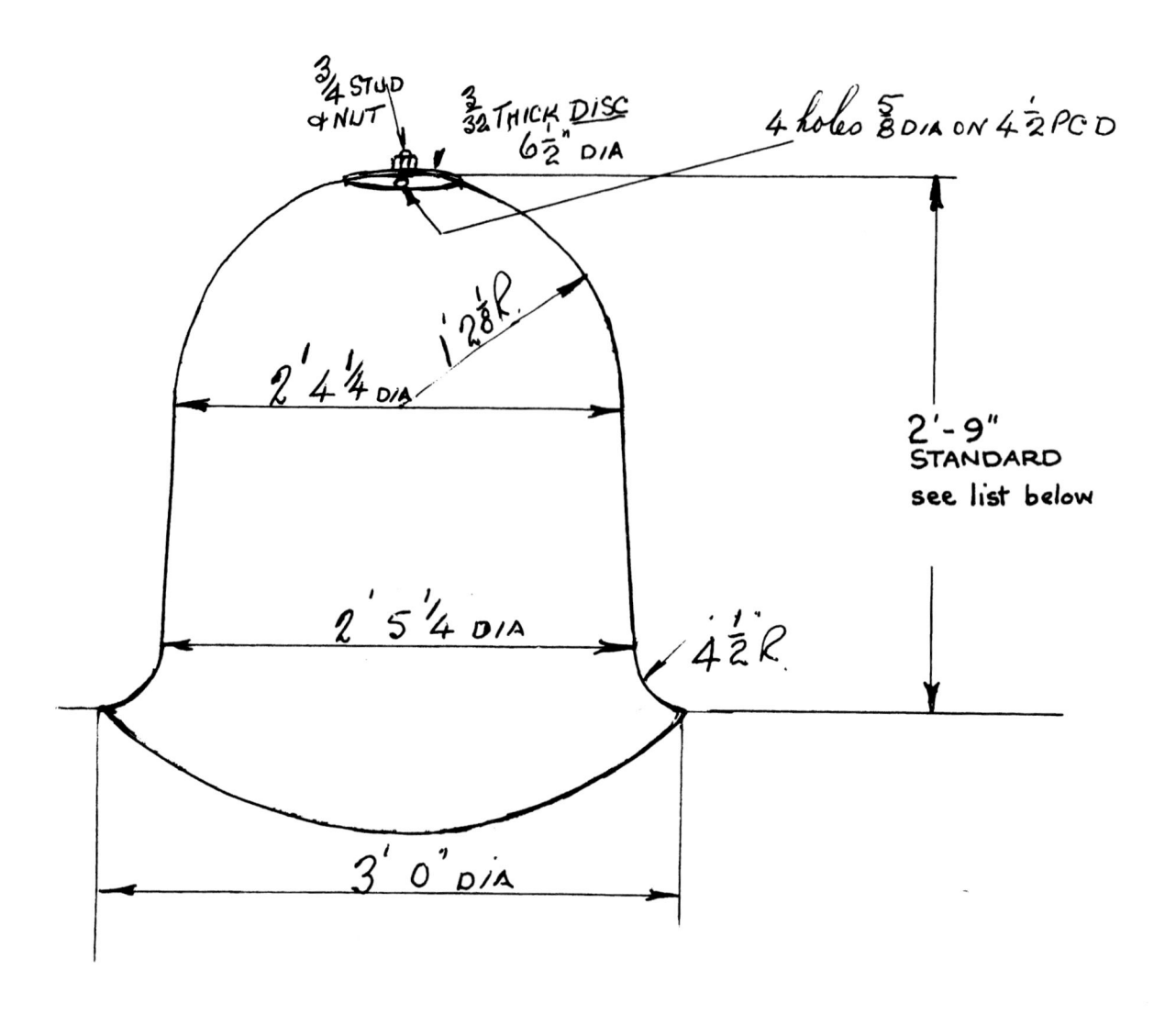

Diagram of steam dome with dimensions:
- ¾" STUD & NUT
- 3/32 THICK DISC 6½" DIA
- 4 holes ⅝" DIA on 4½" PCD
- 1' 2⅛" R
- 2' 4¼" DIA
- 2' 5¼" DIA
- 4½" R
- 3' 0" DIA
- 2'-9" STANDARD see list below

HEIGHT VARIED BETWEEN :

	WITH CHIMNEYS	
2'-8" and 2'-9½"	" "	2'-10" OR LONGER
2'-4"	" "	2'-5"
2'-2"	" "	2'-0"
2'-0"	" "	1'-10"
1'-10½"	" "	1'-8"

TOP OF SMOKEBOX JOINT FLUSH WELDED.

JOINTS ON LAGGING RIVETED — HEADS ALMOST FLUSH WITH JOINT ON TOP OF BOILER.

SAFETY VALVE

Perforated panel each side to allow condensate to escape

later shortened to outside of cab

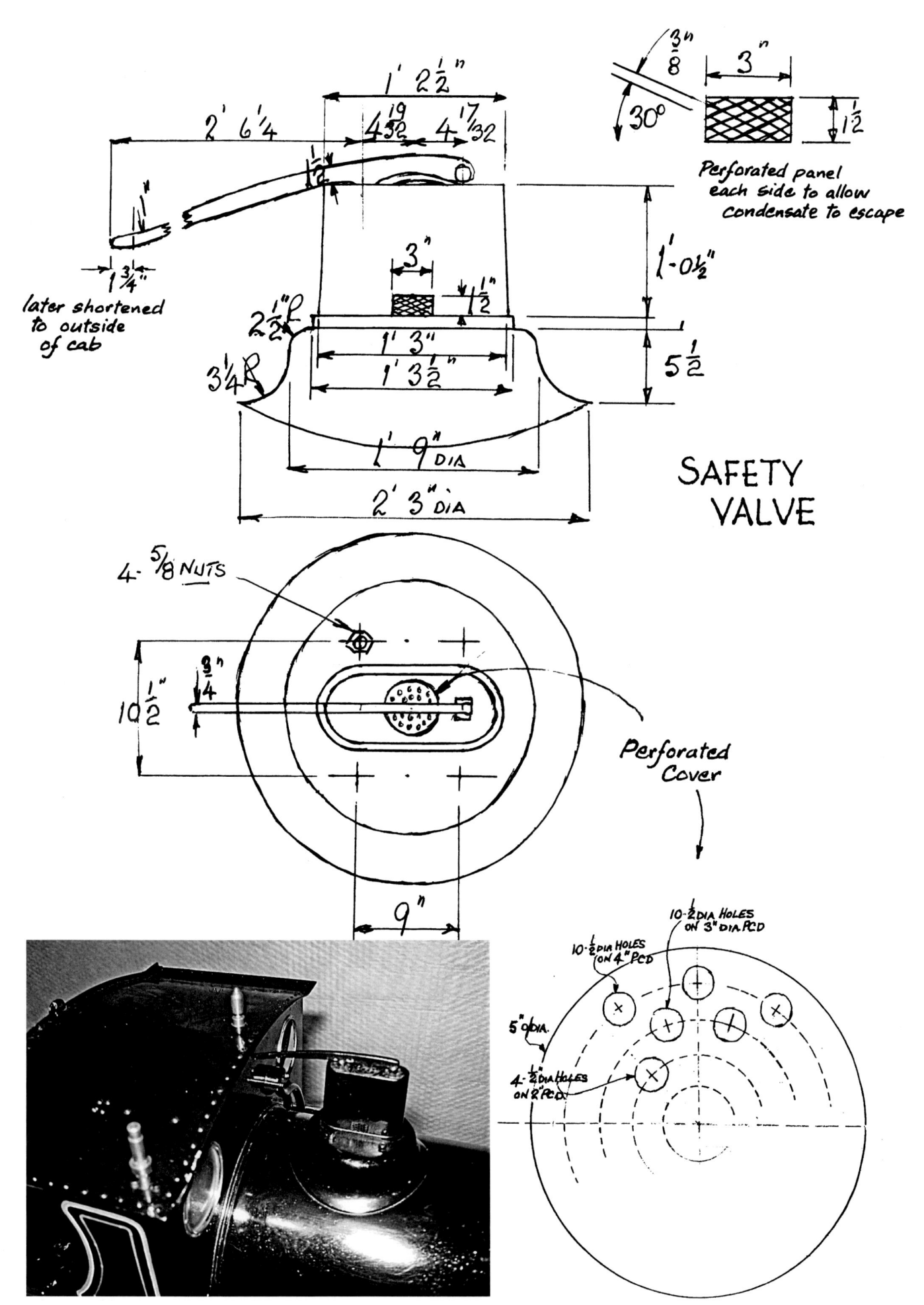

4 - 5/8" NUTS

Perforated Cover

10 - 1/2" DIA HOLES ON 4" PCD
10 - 2" DIA HOLES ON 3" DIA PCD
4 - 1/2" DIA HOLES ON 2" PCD
5" DIA.

Simon Fountain

Smokebox Regulator Lubricator

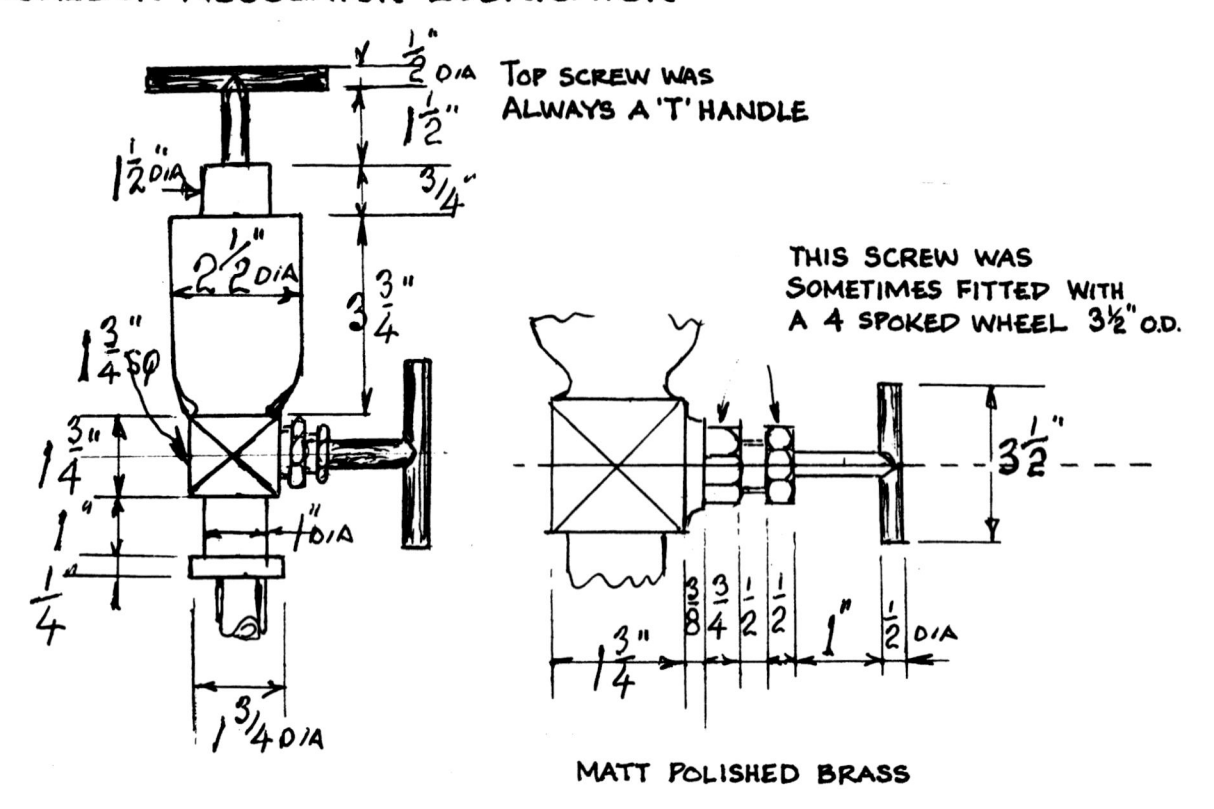

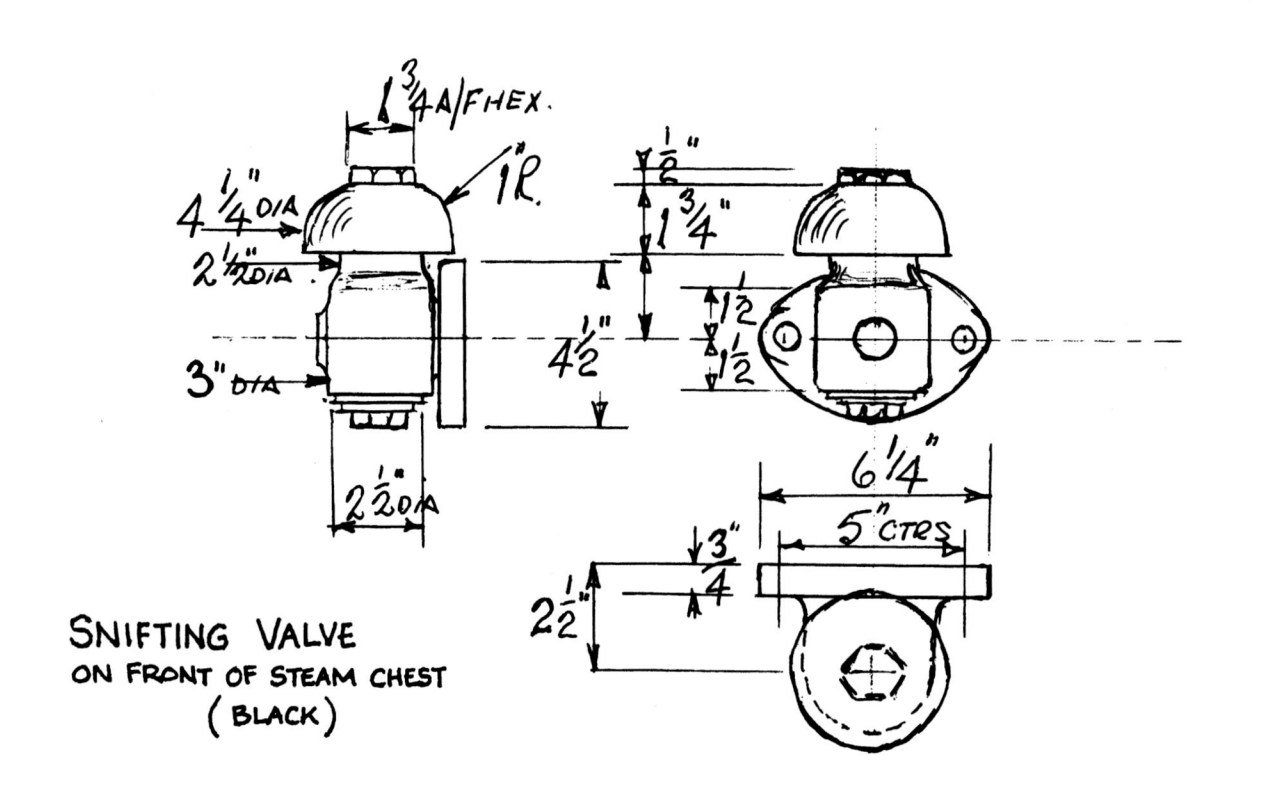

Top screw was always a 'T' handle

This screw was sometimes fitted with a 4 spoked wheel 3½" O.D.

MATT POLISHED BRASS

Snifting Valve
on front of steam chest
(Black)

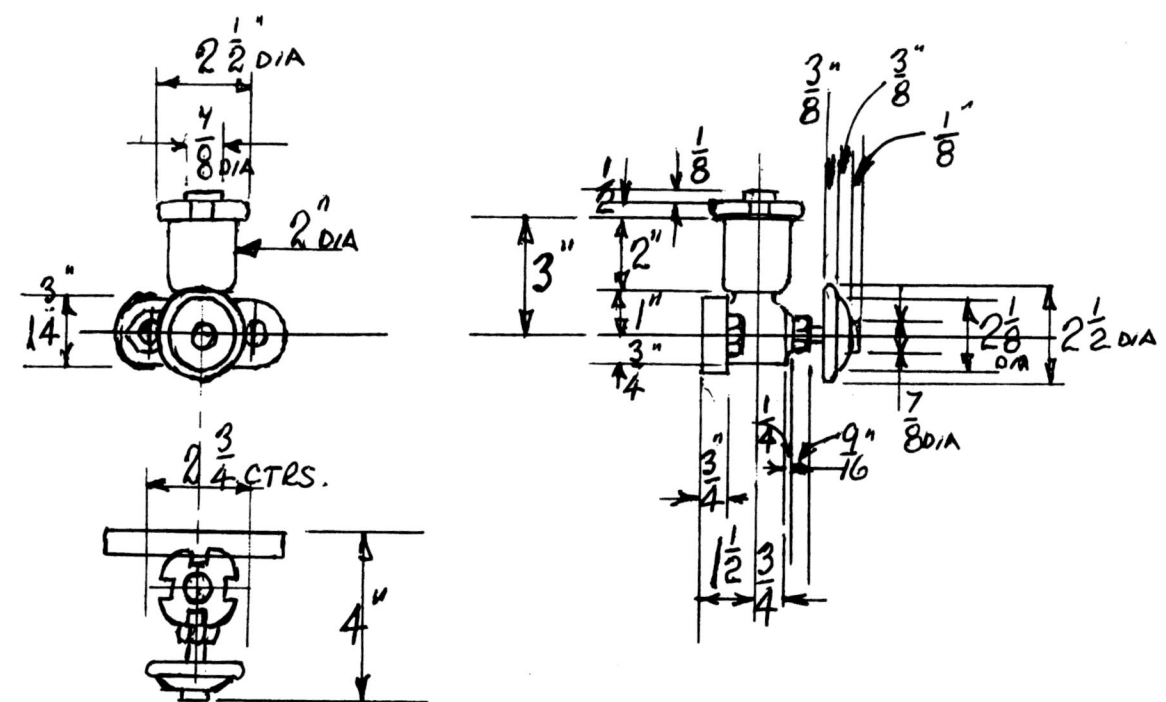

LUBRICATOR CUPS ON CYLINDER COVERS

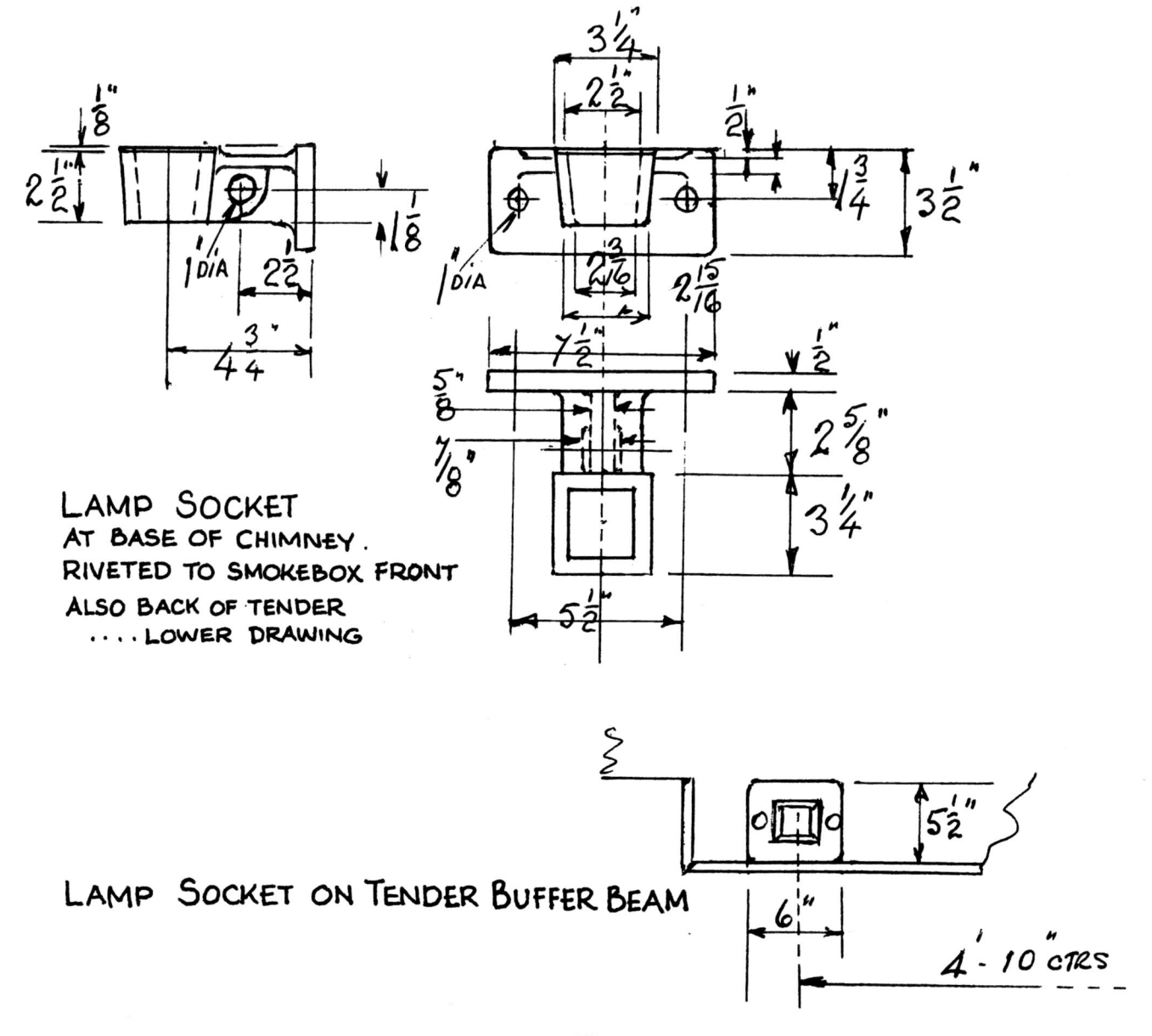

LAMP SOCKET
AT BASE OF CHIMNEY.
RIVETED TO SMOKEBOX FRONT
ALSO BACK OF TENDER
....LOWER DRAWING

LAMP SOCKET ON TENDER BUFFER BEAM

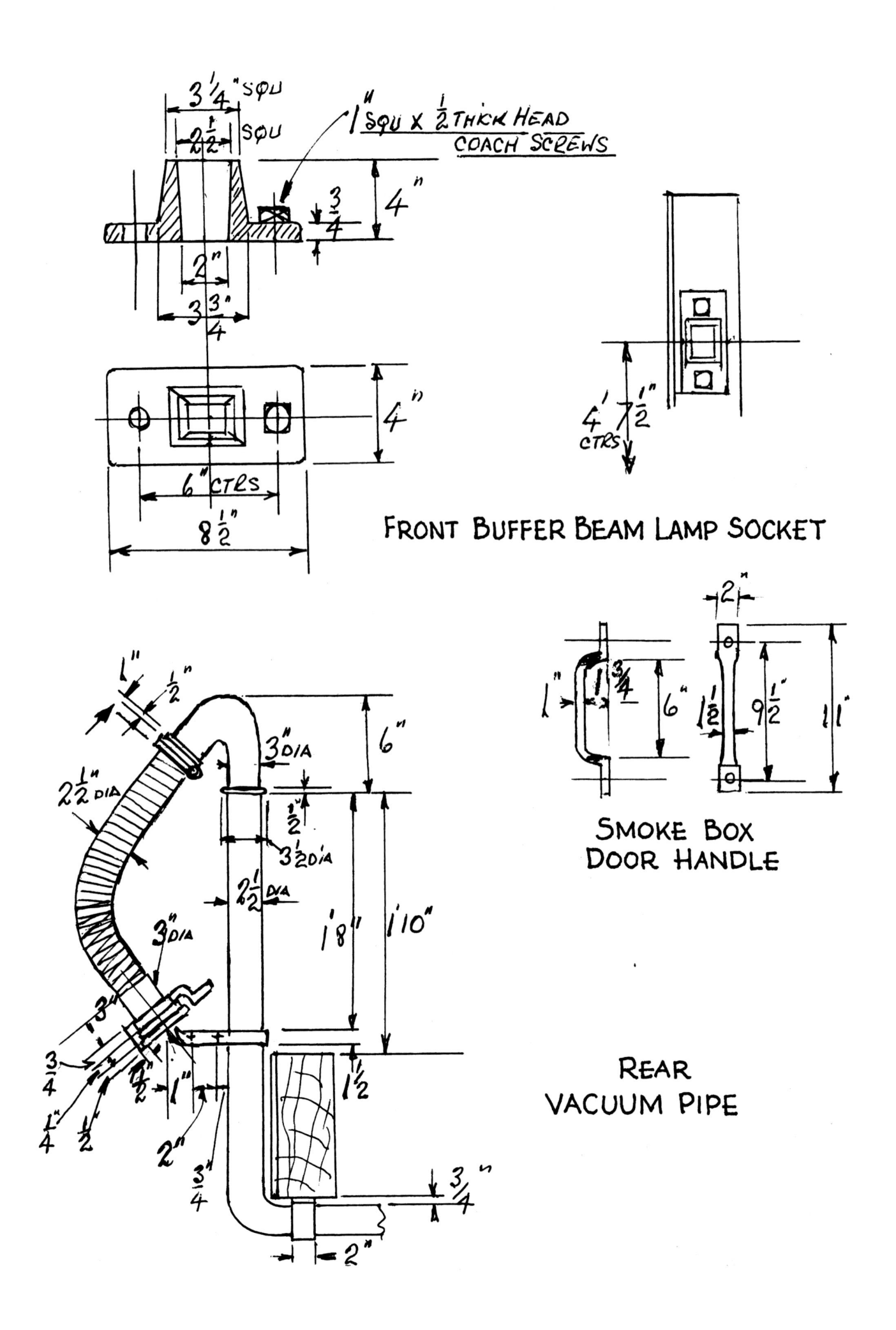

FRONT BUFFER BEAM LAMP SOCKET

SMOKE BOX DOOR HANDLE

REAR VACUUM PIPE

Front vacuum pipe

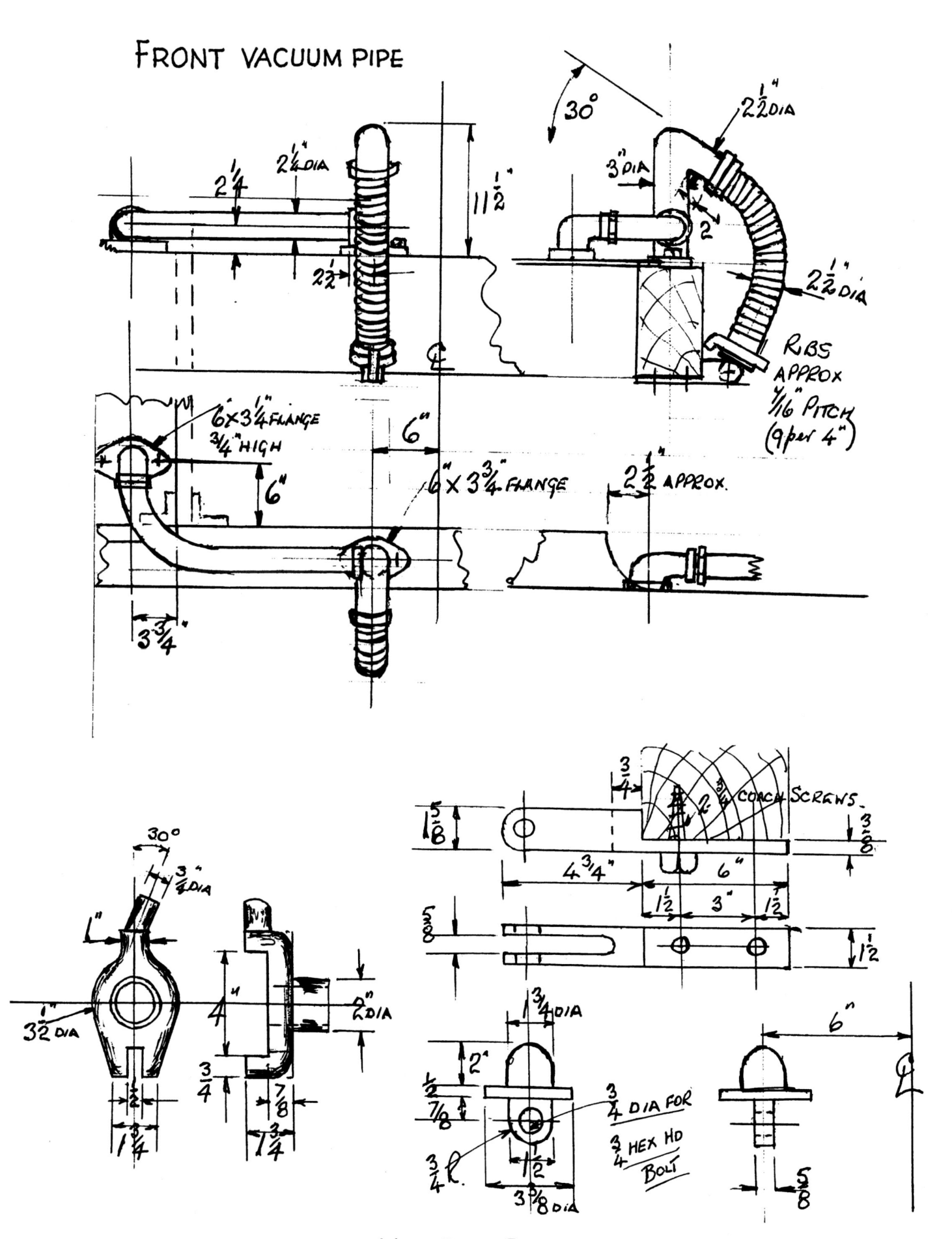

Vac Pipe Bracket Under Front Buffer Beam

SMOKE BOX DOOR HINGE

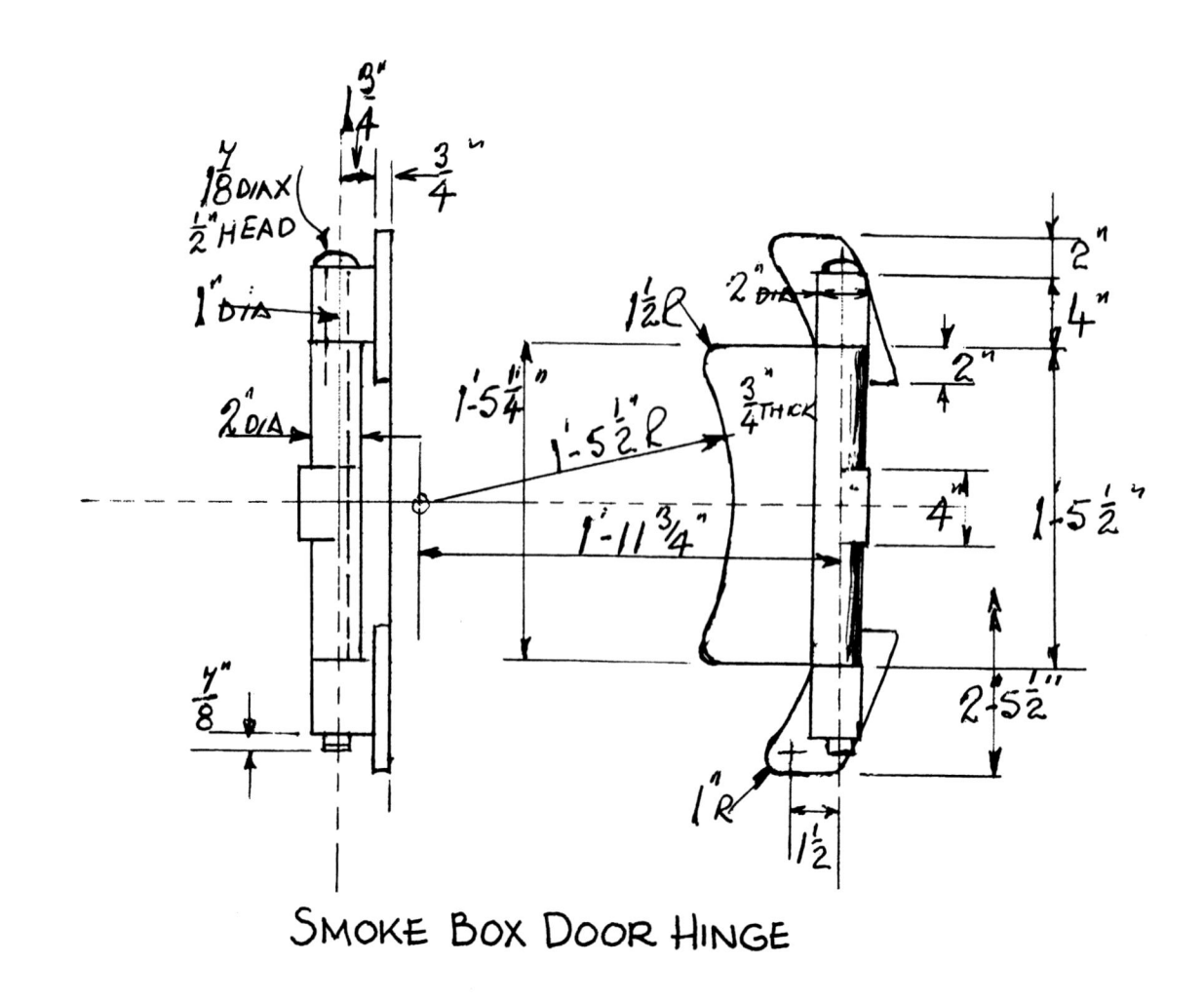

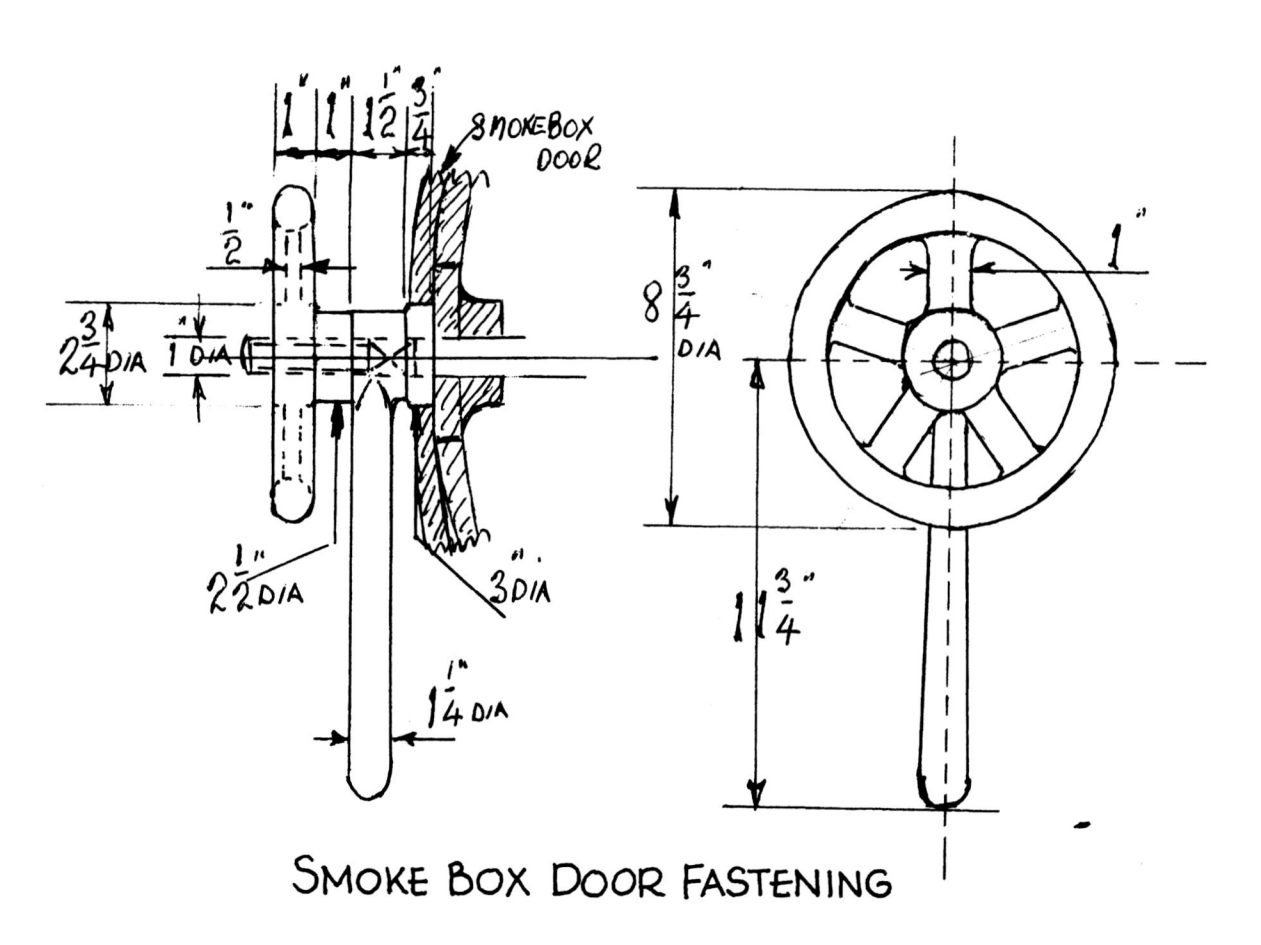

SMOKE BOX DOOR FASTENING

Handrail Knobs
Smokebox Front

2" DIA
1¼" DIA
2" DIA
1⅛" DIA HOLE
2½"
¼"

Smokebox Dogs

2"
¾"
1½"
2"
1¼"
1"
1" DIA
2' 6"

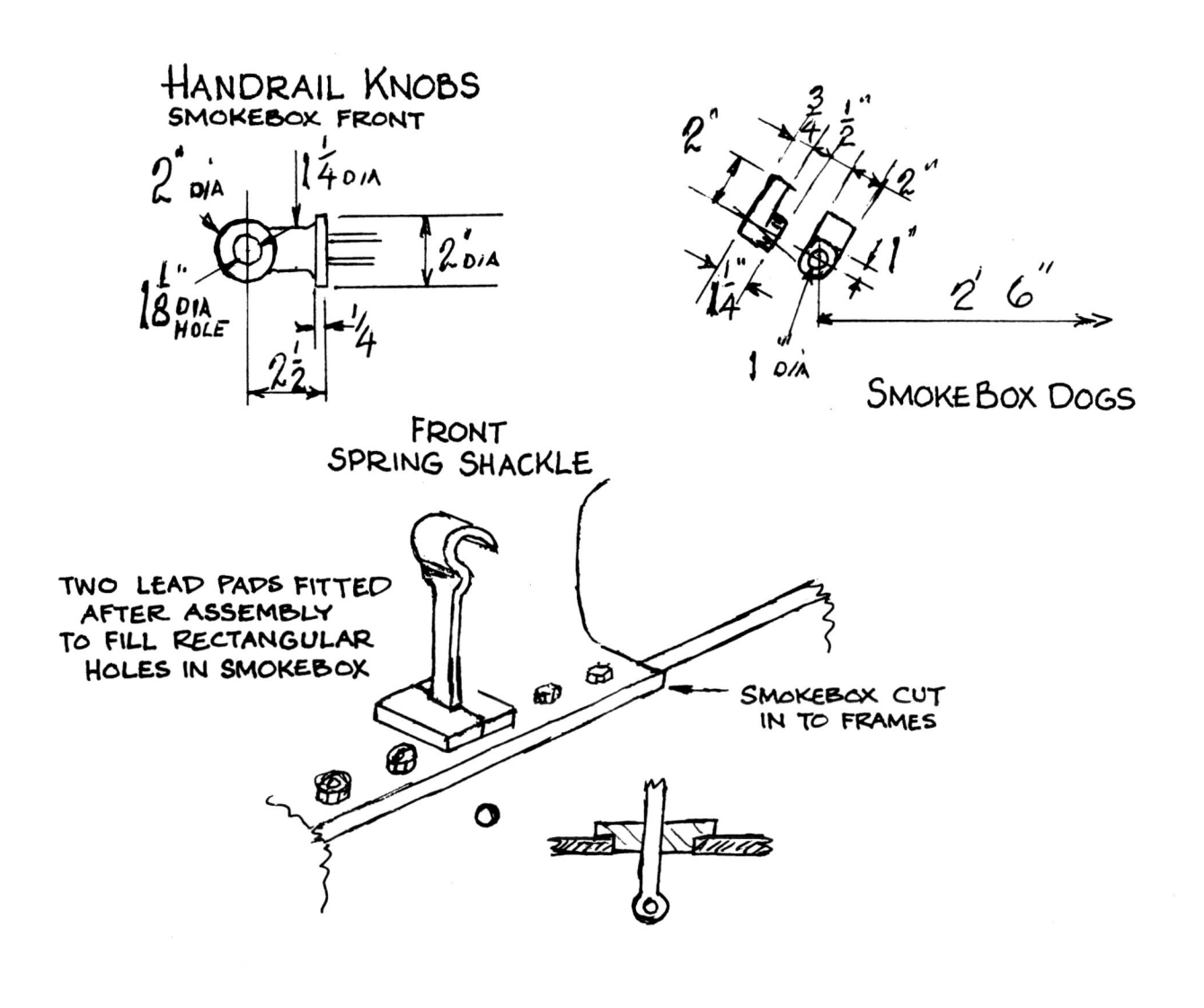

Front Spring Shackle

TWO LEAD PADS FITTED AFTER ASSEMBLY TO FILL RECTANGULAR HOLES IN SMOKEBOX

SMOKEBOX CUT IN TO FRAMES

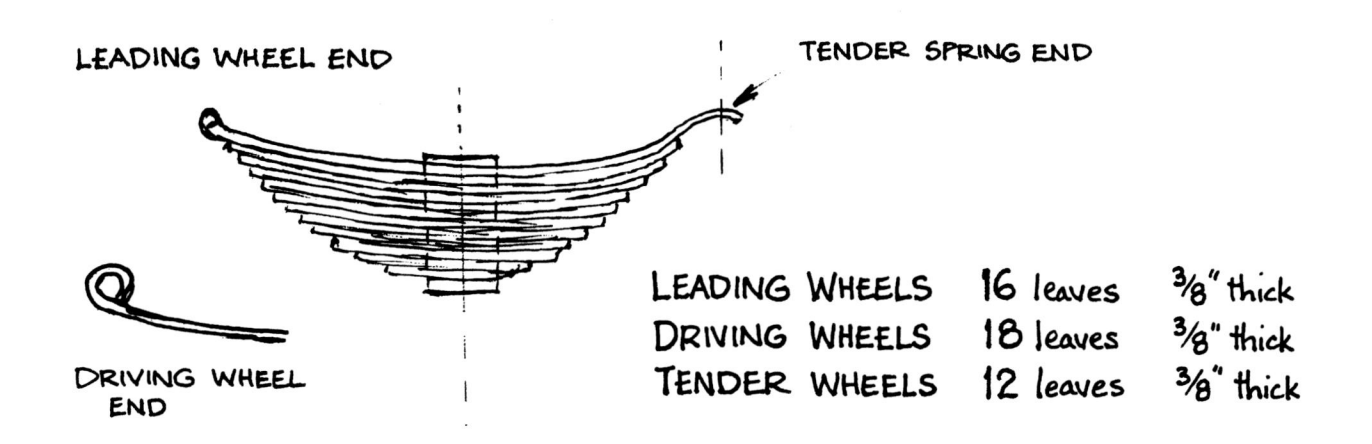

LEADING WHEEL END

TENDER SPRING END

DRIVING WHEEL END

LEADING WHEELS	16 leaves	⅜" thick
DRIVING WHEELS	18 leaves	⅜" thick
TENDER WHEELS	12 leaves	⅜" thick

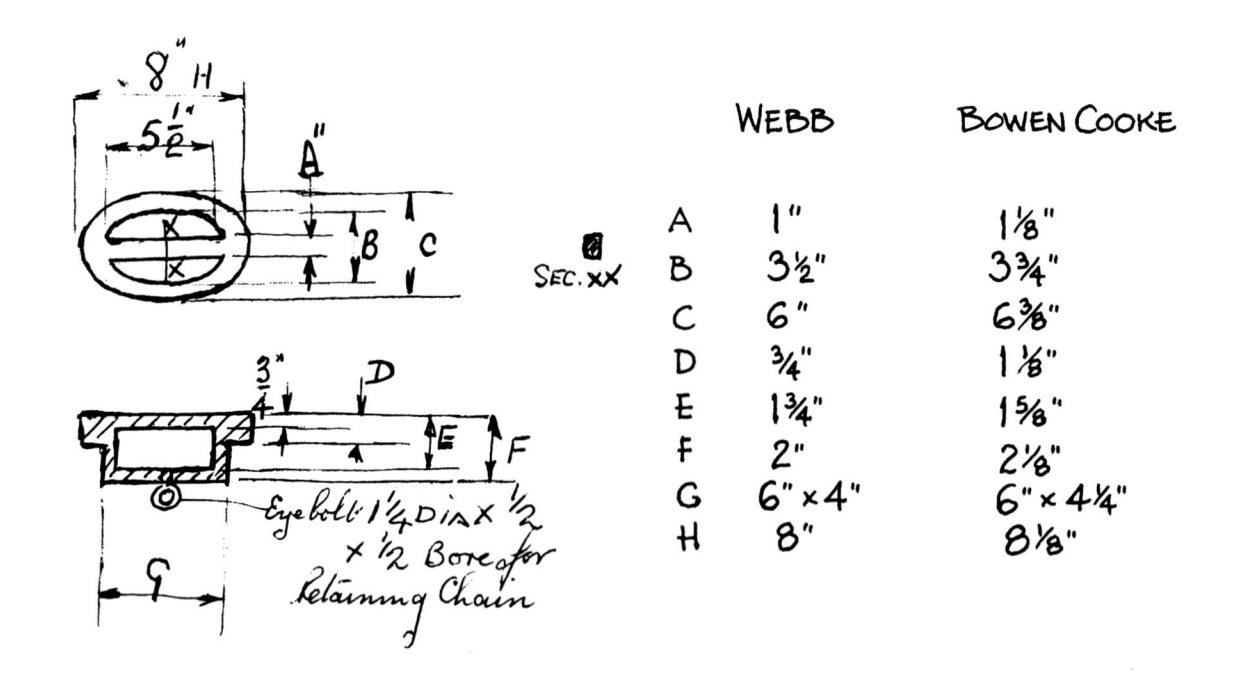

SAND BOX COVERS

	WEBB	BOWEN COOKE
A	1"	1⅛"
B	3½"	3¾"
C	6"	6⅜"
D	¾"	1⅛"
E	1¾"	1⅝"
F	2"	2⅛"
G	6" × 4"	6" × 4¼"
H	8"	8⅛"

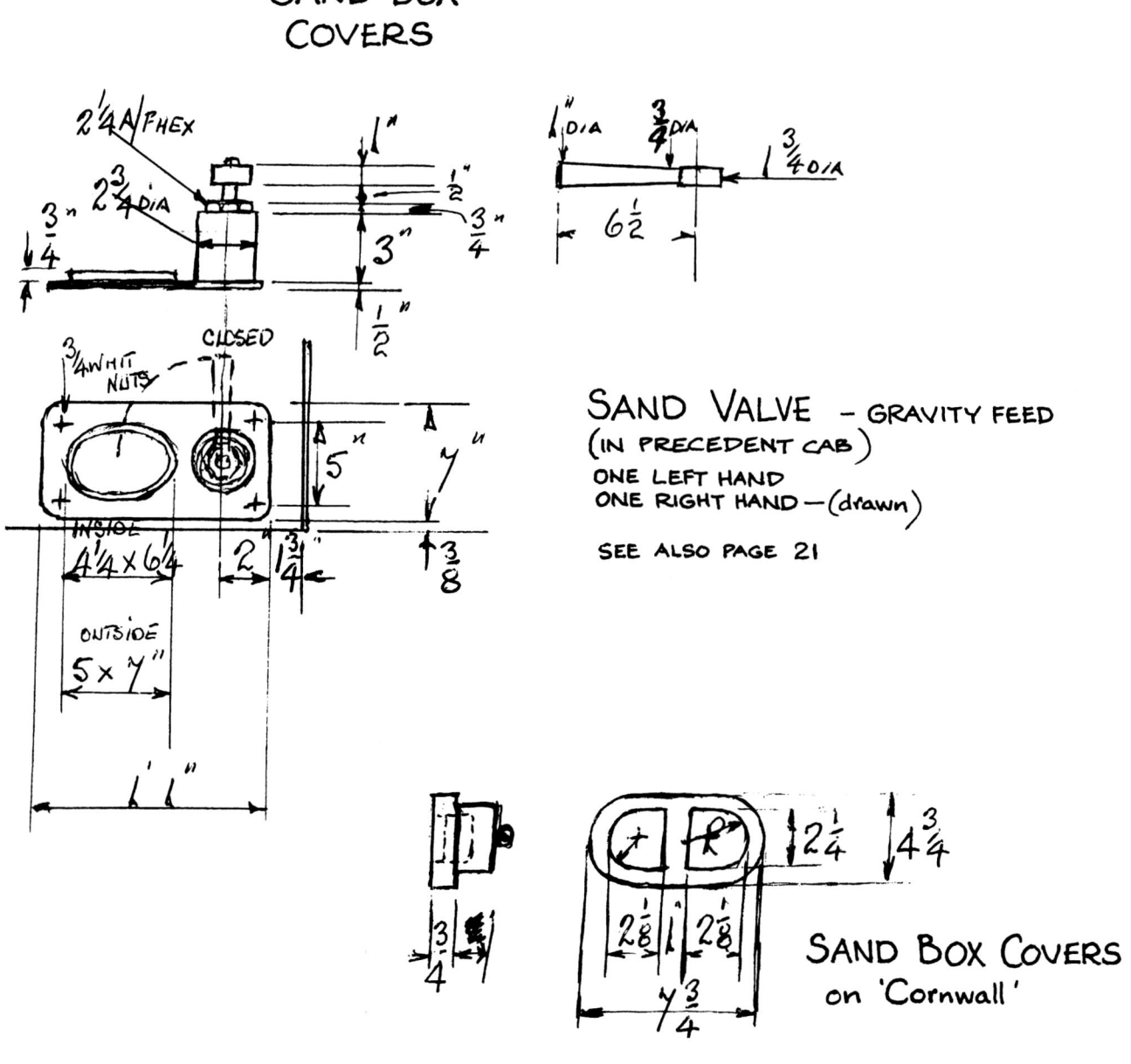

SAND VALVE – GRAVITY FEED
(IN PRECEDENT CAB)
ONE LEFT HAND
ONE RIGHT HAND – (drawn)

SEE ALSO PAGE 21

SAND BOX COVERS
on 'Cornwall'

SAFETY HOOKS
ON TENDER

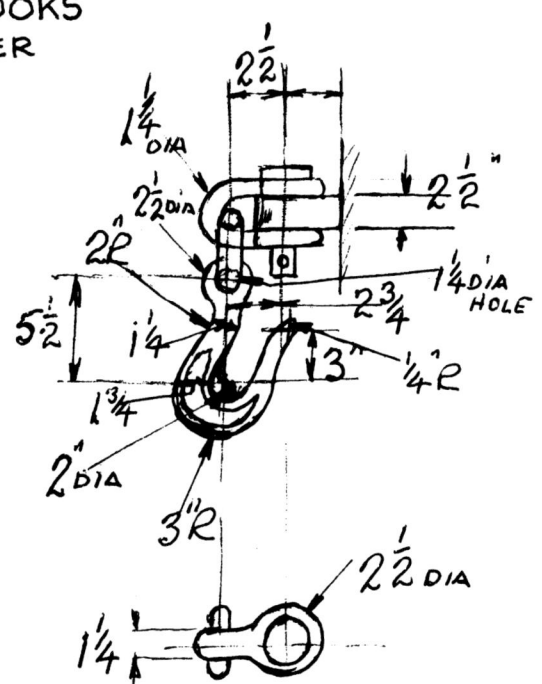

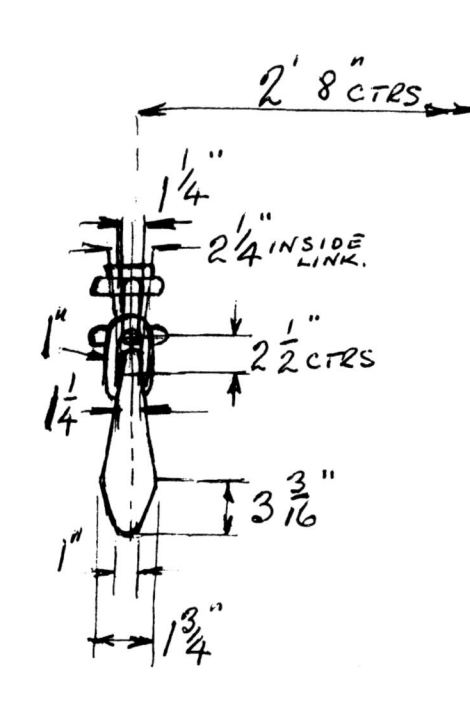

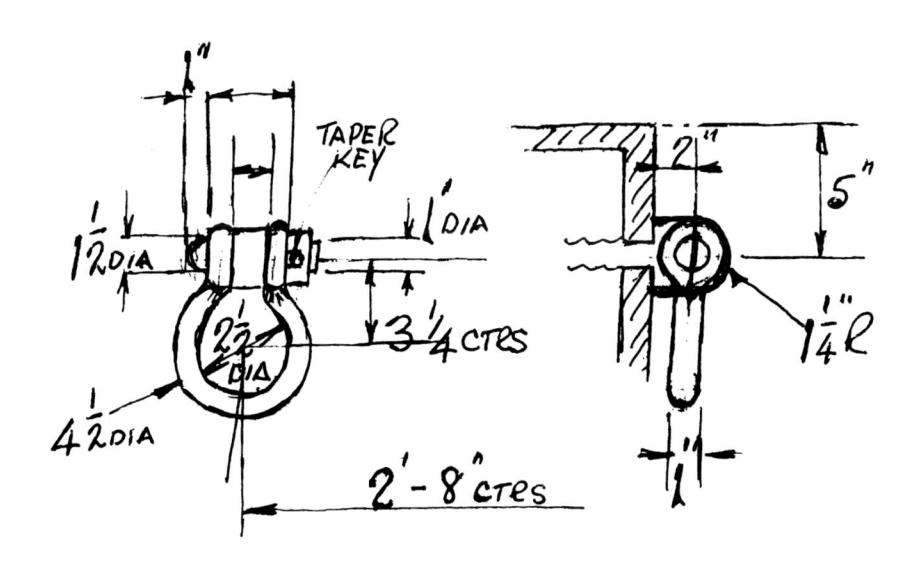

SAFETY HOOKS
ON DRAG BOX

BOILER BANDS ··· 2" WIDE

SIZE OF RIVET HEADS
    CAB . . . . 3/4" dia.
    SMOKEBOX . . . 3/4" dia.
    TOOL BOXES . . 1/2" dia.
    TENDER . . . . . 7/8" dia.

STAY NUTS ON BACKHEAD 1 7/8" AF HEX. × 1"
BOLTS (NUTS) ON CYLINDER COVERS 1 3/8" AF HEX.
HEADS OF BOLTS IN SPRING SHACKLES
    AND GUARD IRONS . . . . . . 1 3/4" dia × 3/8" thick

HEX. BOLTS IN WHEEL TYRES 2 1/8" AF HEX. × 7/32" thick

SPOKES OF WHEELS — RECTANGULAR (2" × 1 1/2") × (4 × 3)

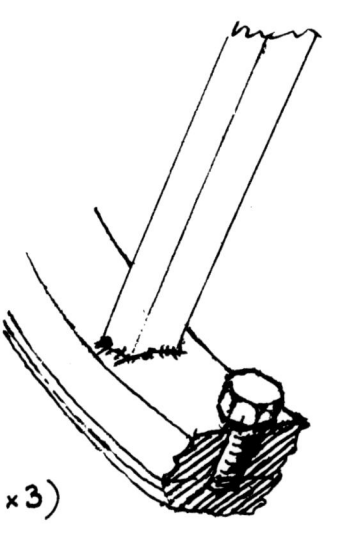

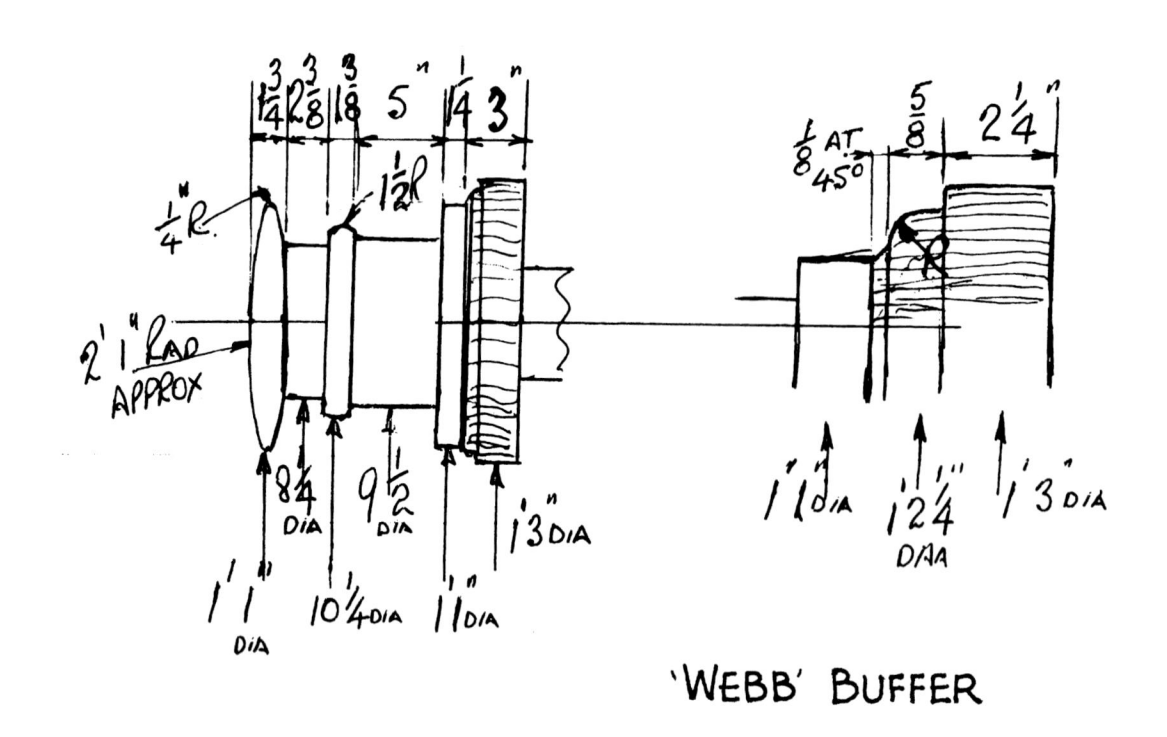

'WEBB' BUFFER

'COOKE' PATTERN BUFFER

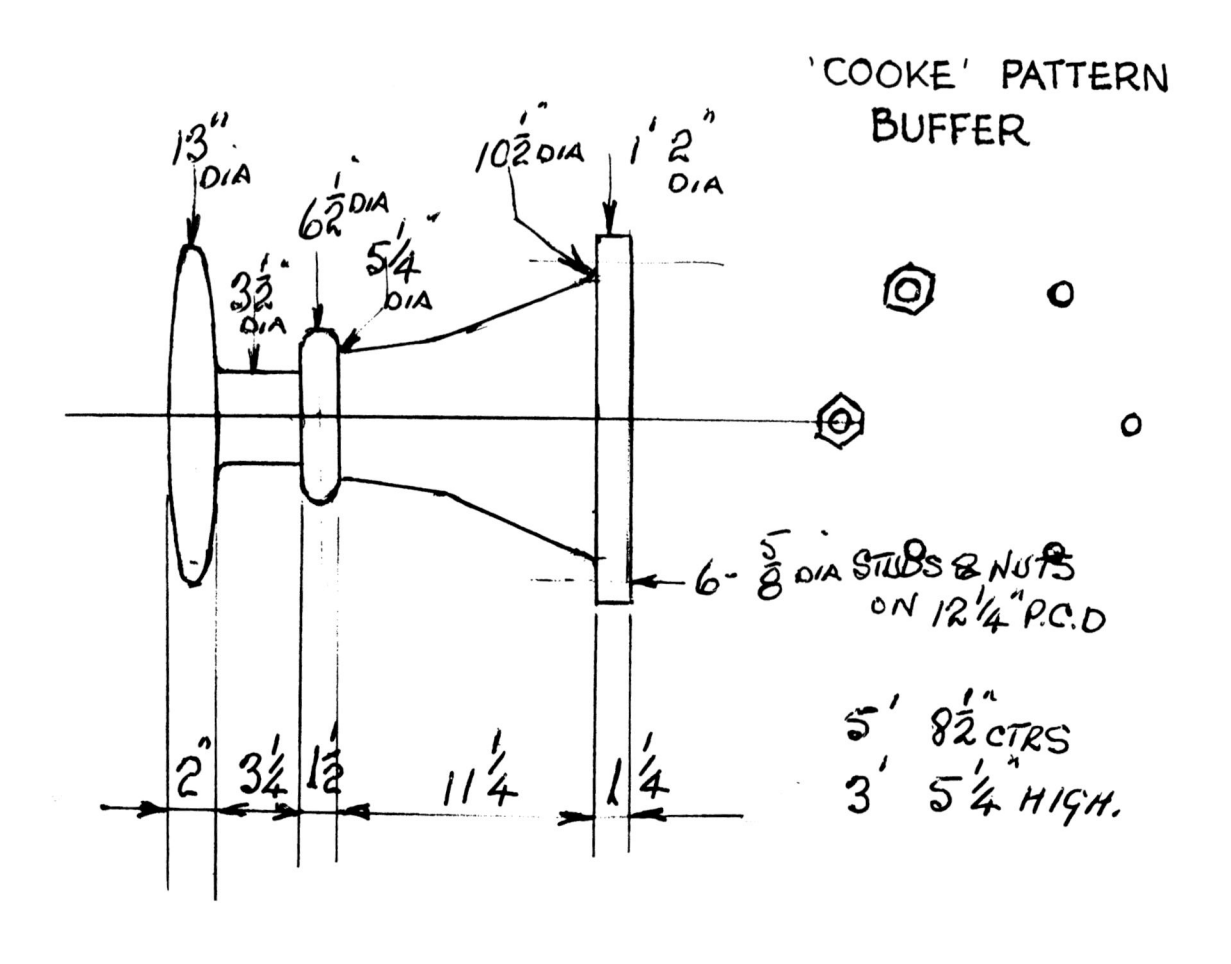

6 - ⅝" DIA STUDS & NUTS ON 12¼" P.C.D

5' 8½" CTRS
3' 5¼" HIGH.

SLIPPERS – STEEL
WHITE METAL FACES

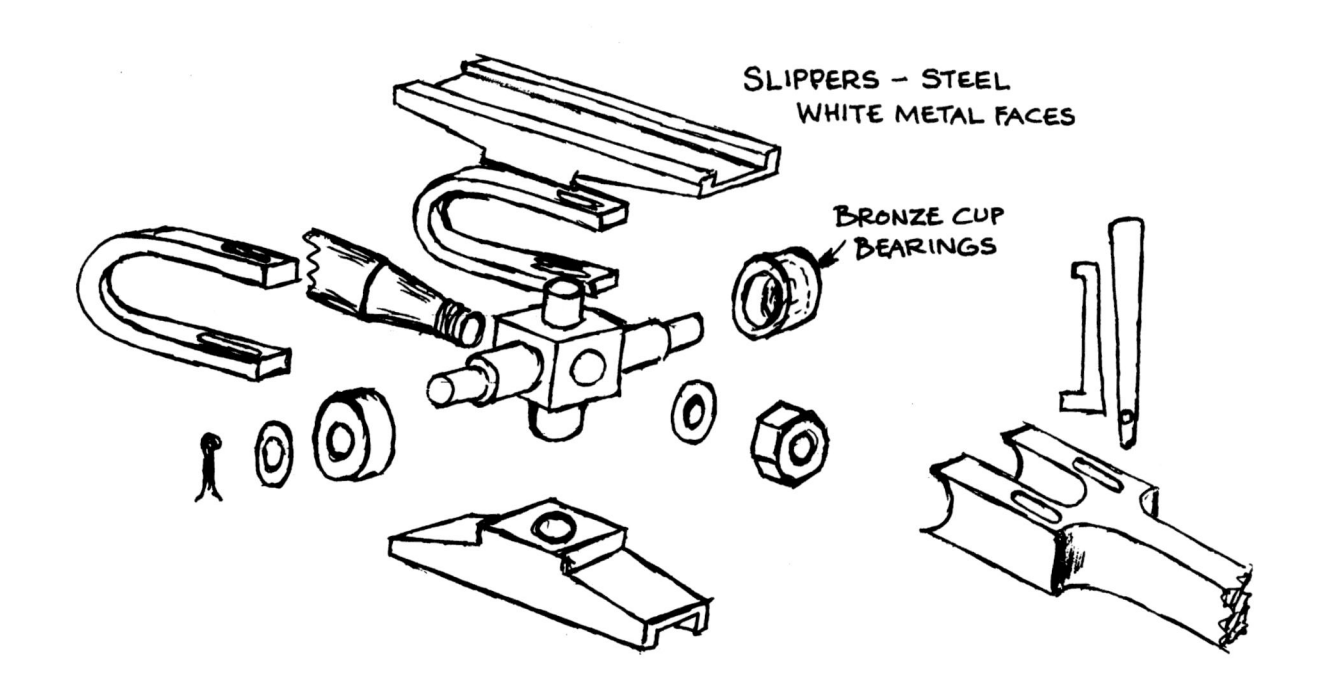

BRONZE CUP BEARINGS

CONN. ROD – Small End

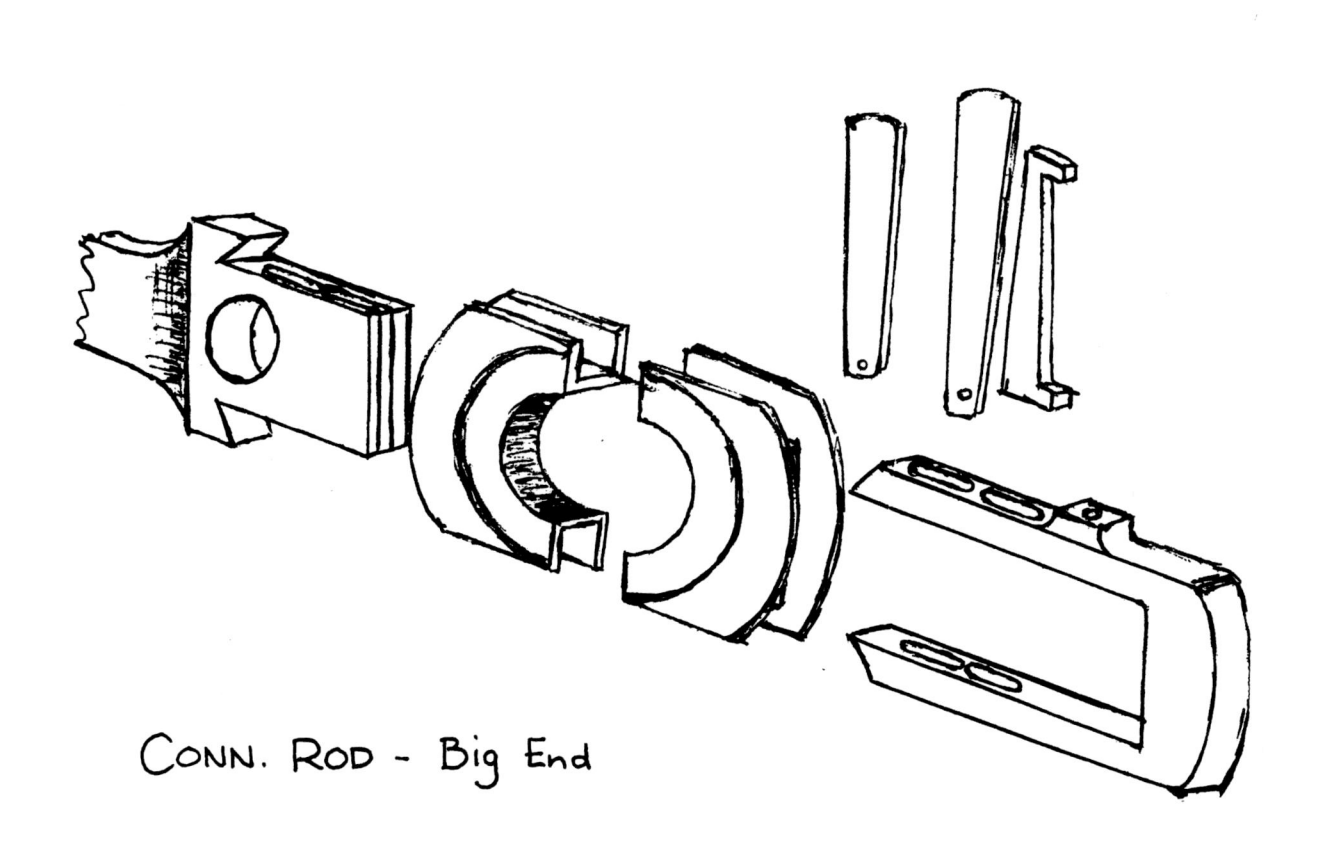

CONN. ROD – Big End

Mike Williams

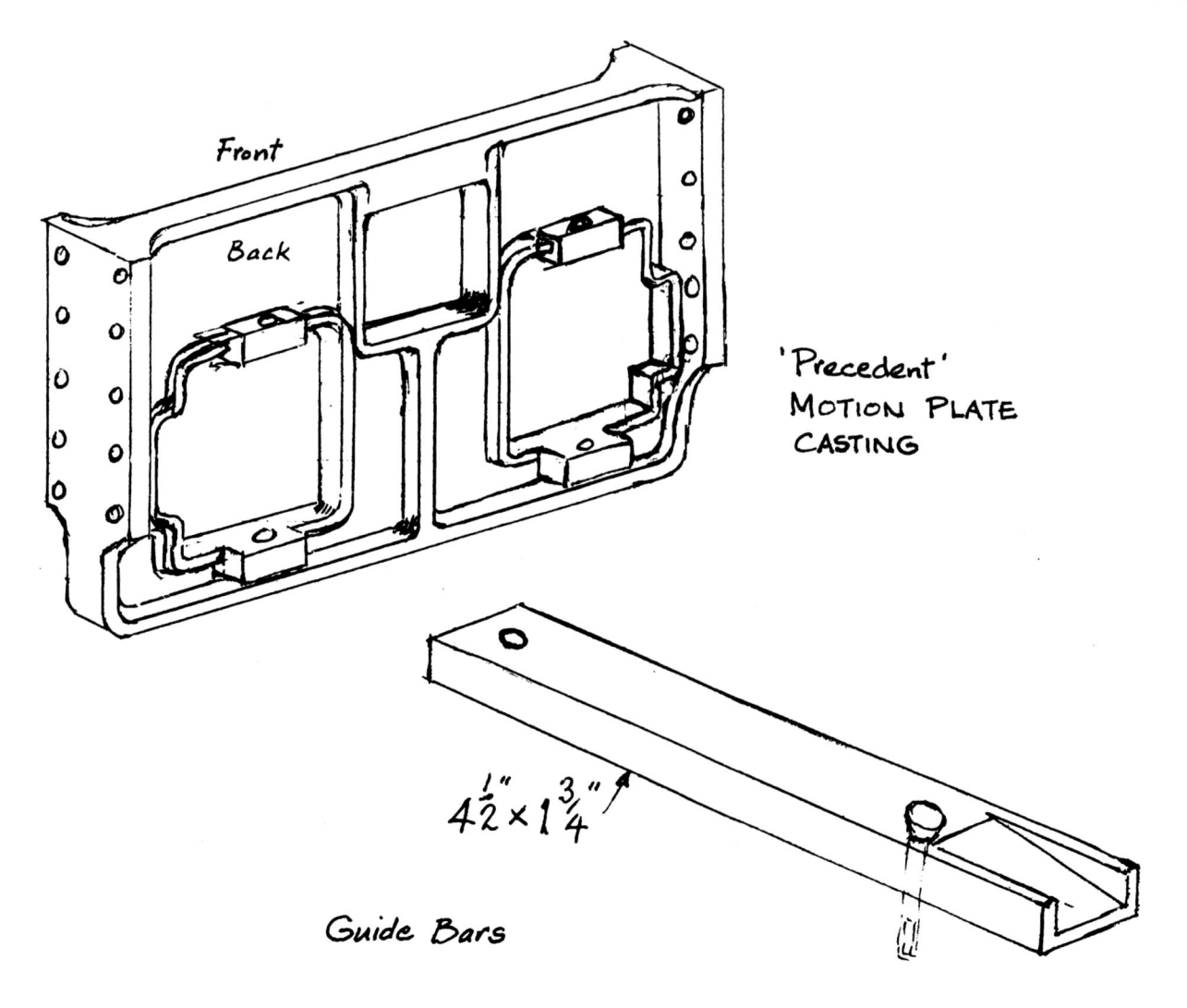

'Precedent' MOTION PLATE CASTING

Front
Back

$4\tfrac{1}{2}" \times 1\tfrac{3}{4}"$

Guide Bars

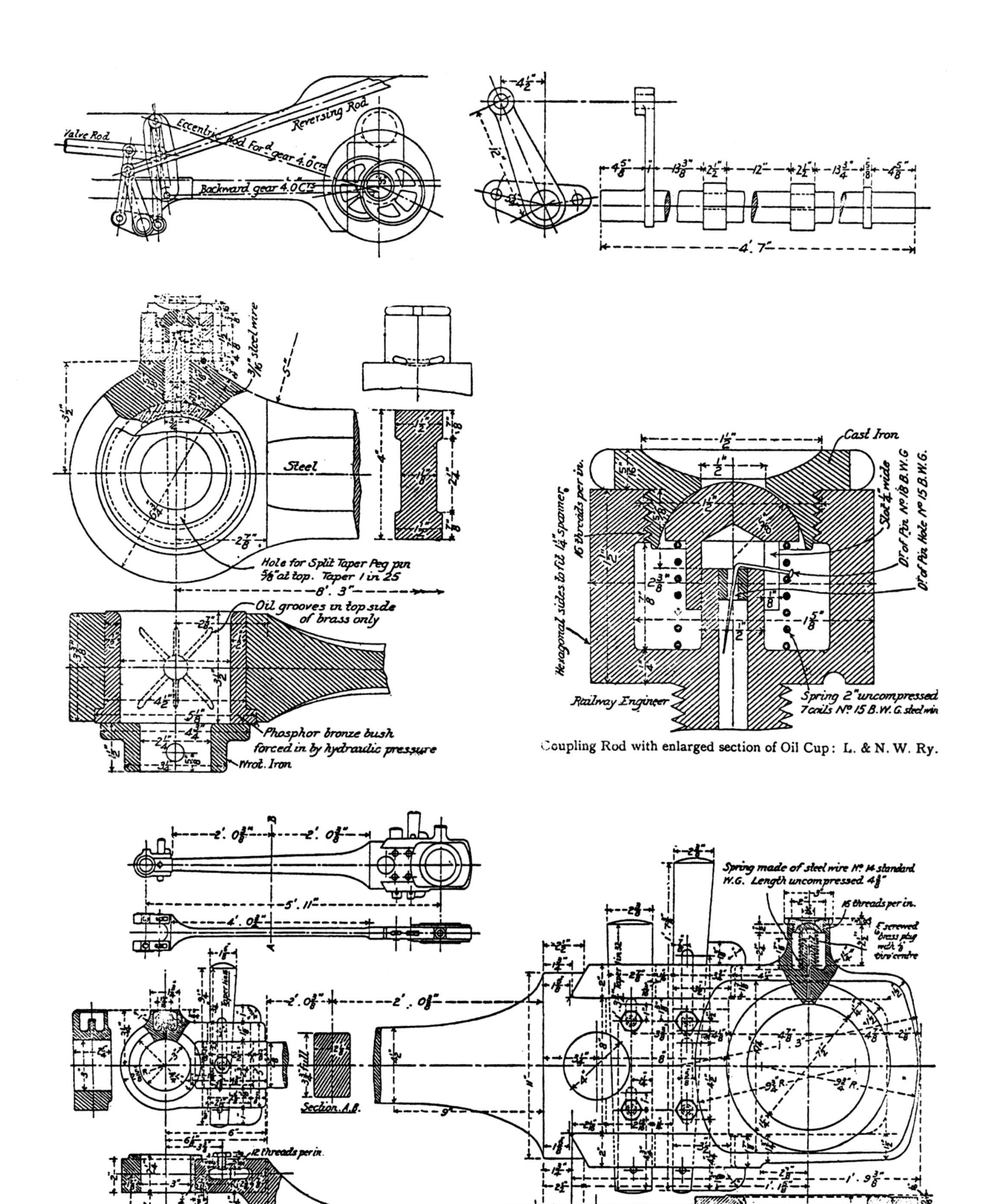

Coupling Rod with enlarged section of Oil Cup: L. & N. W. Ry.

Standard Connecting Rod: London and North Western Railway; Interchangeable on 2,300 engines.

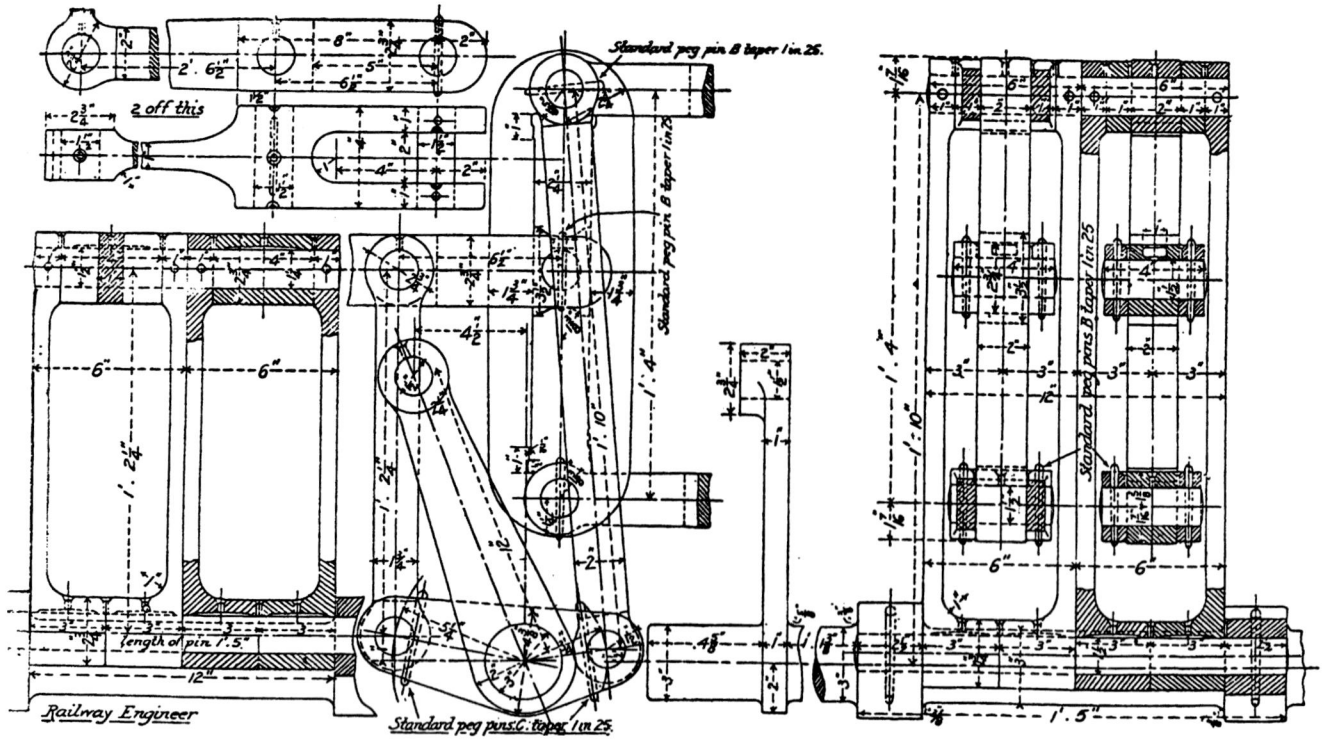

Standard Valve Motion for 17" × 24" Engines: London and North Western Railway.

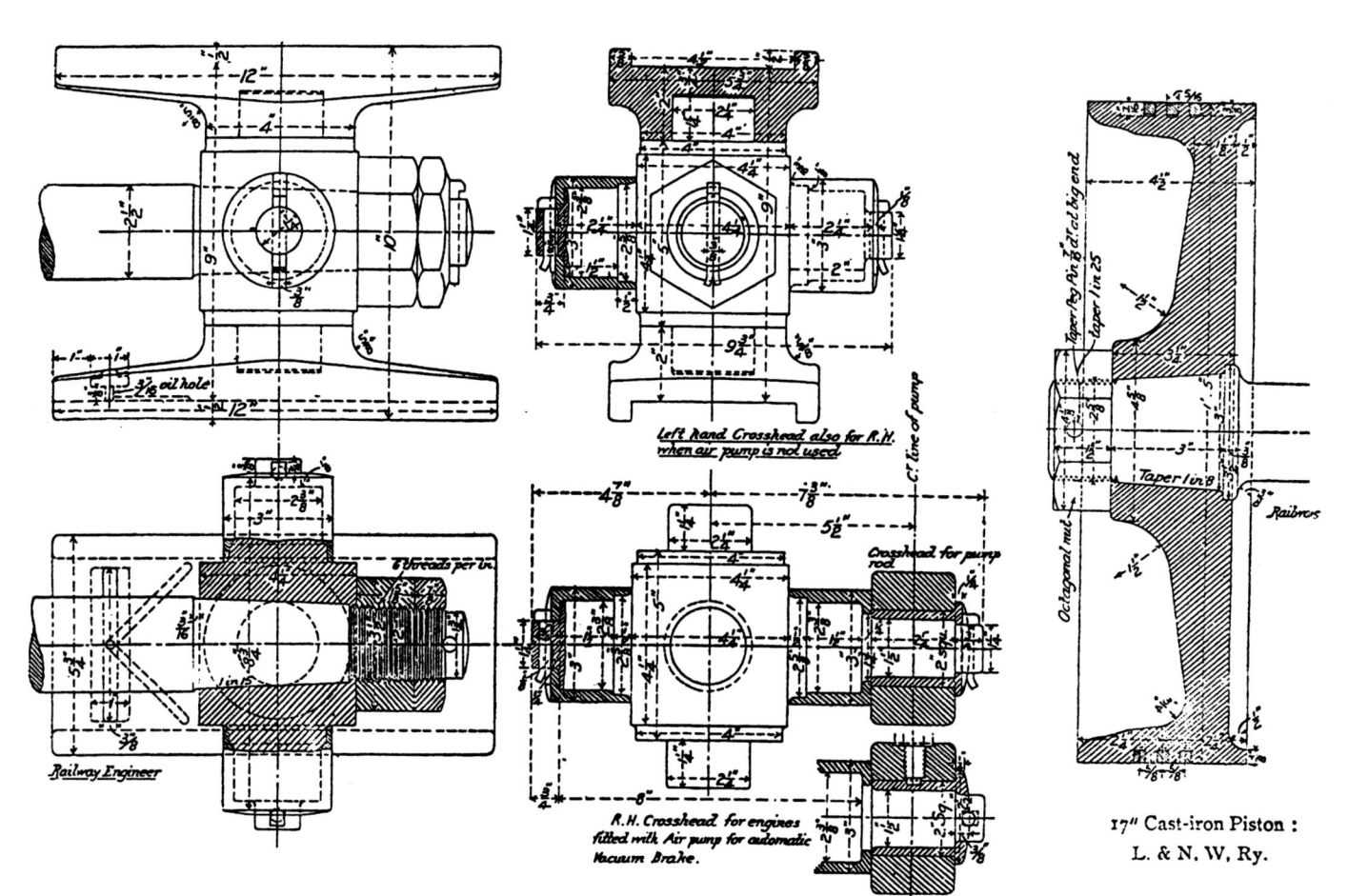

Standard Cross-head 17" × 24" Engines: London and North Western Railway.

17" Cast-iron Piston: L. & N. W. Ry.

VACUUM EJECTOR PIPE SUPPORT

- 5"
- Lock Nut
- Nut
- 1¼" A/F
- 4" Rad.
- ½"
- 10"
- Radius to suit Splasher
- 3½"
- 1¾" Radius
- Height to suit Vac. Ejector Pipe
- 5½"

HAND RAIL KNOBS

- 1" DIA
- 1¾" DIA
- 1" DIA HAND RAIL
- 1/8" DIA HOLE
- 2" DIA. SPH.
- Vacuum Ejector Pipe 3½" DIA
- ⅝"
- 1½"
- ¾"
- 2" Boiler Band
- 1¼"

Knobs in line with Boiler Bands

OIL BOXES (Precedents)

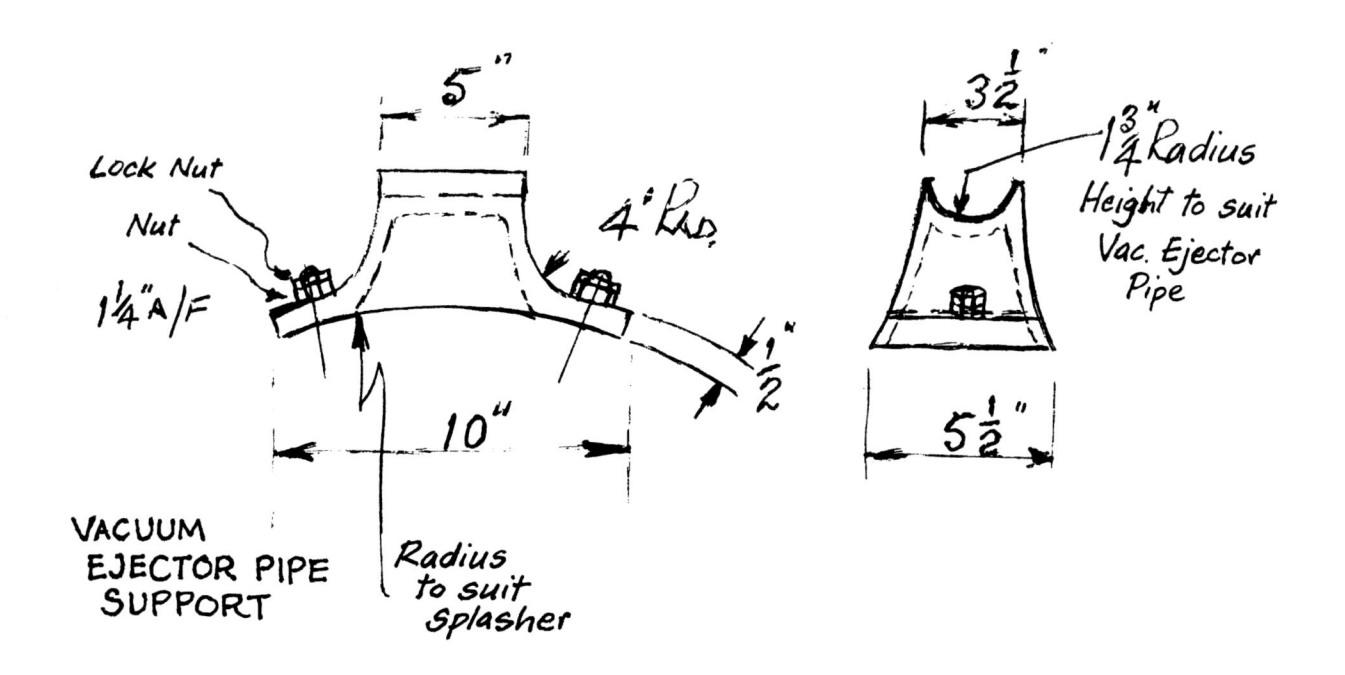

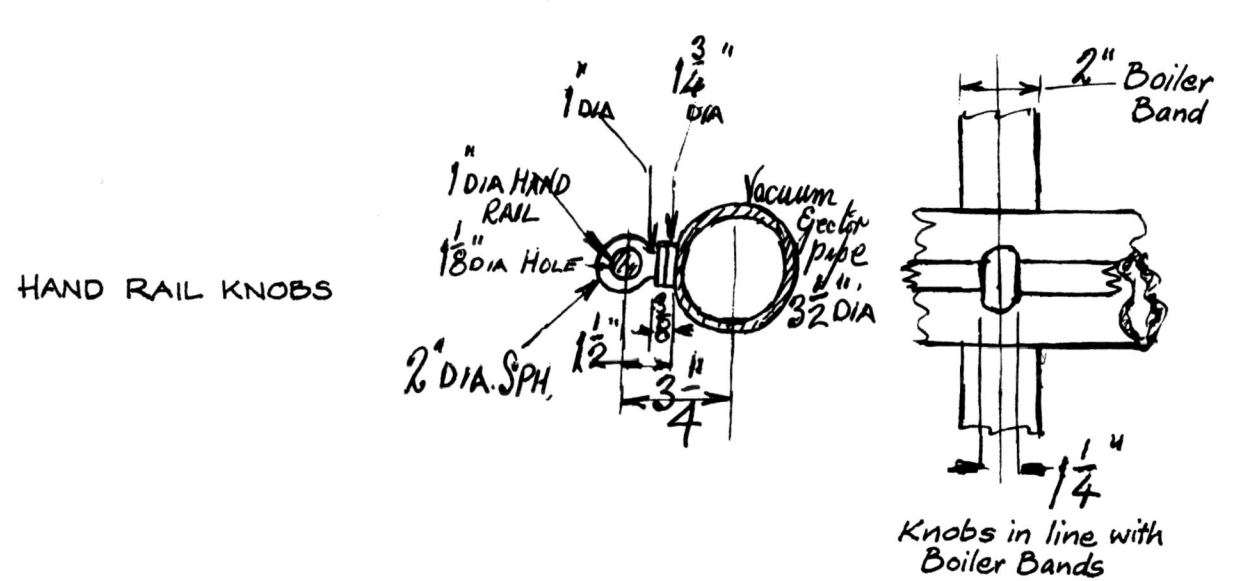

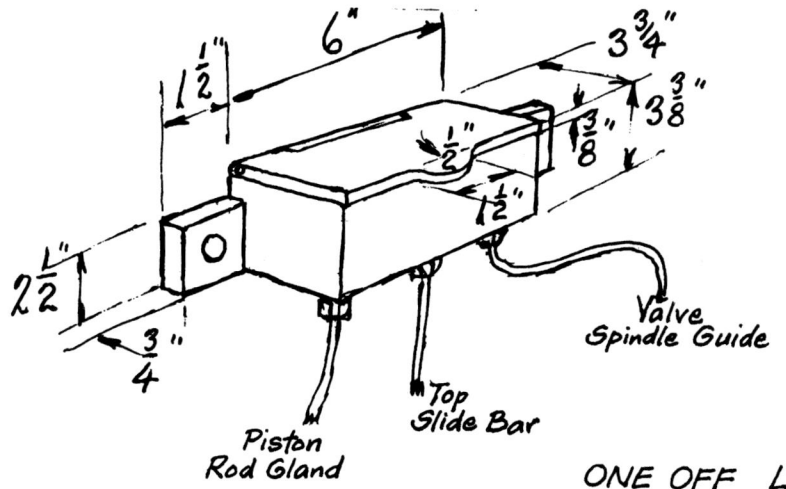

- 6"
- 1½"
- 3¾"
- 3⅜"
- ½"
- ½"
- ⅜"
- 2½"
- ¾"
- Piston Rod Gland
- Top Slide Bar
- Valve Spindle Guide

LID : polished brass
BODY : painted black
COPPER PIPES : polished

ONE OFF   Left Hand as shown.
ONE OFF   Right Hand on back face of steam chest.
ONE OFF   Each side inside of frames behind smoke box to leading axle boxes.
ONE OFF   Each side, inside of frames near floor of footplate to rear axleboxes.

Fixed to underneath of Drag Box

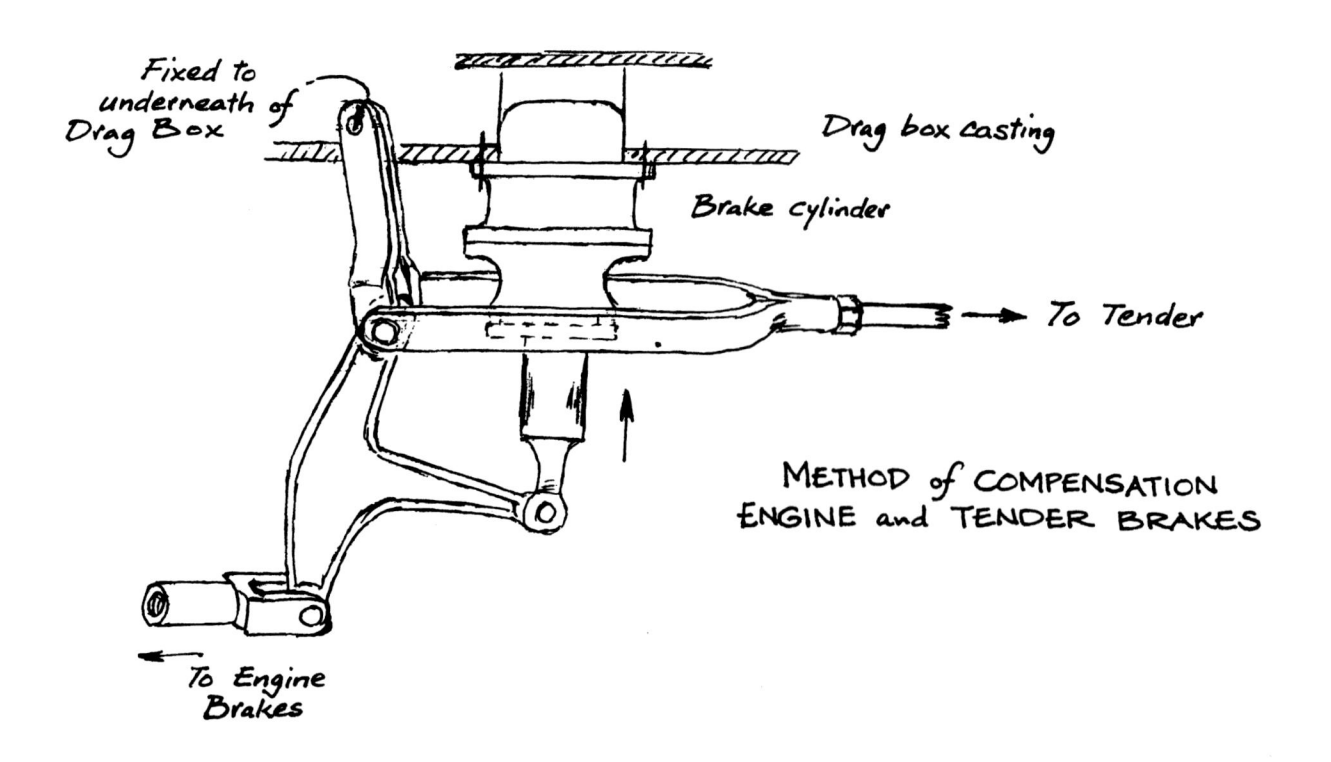

Drag box casting

Brake cylinder

→ To Tender

← To Engine Brakes

METHOD of COMPENSATION ENGINE and TENDER BRAKES

BRAKE GEAR COMPENSATION BETWEEN THE DRIVING WHEELS

6'-9" Precendents
5'-6" 2-4-0 Precursors
6'-6" Newtons
2-2-2-2 Experiments

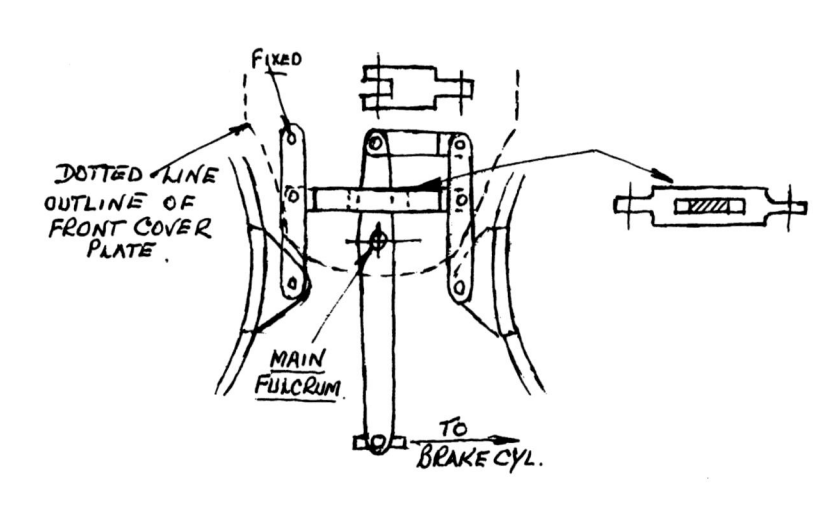

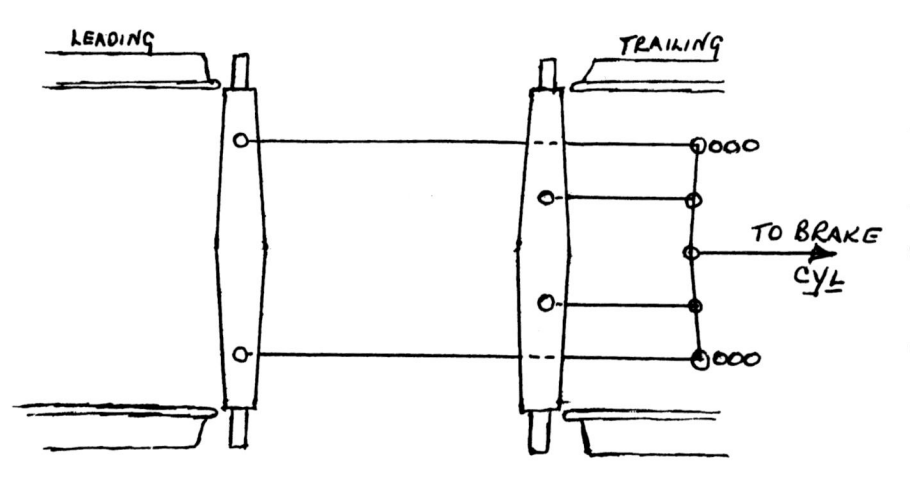

FIXED

DOTTED LINE OUTLINE OF FRONT COVER PLATE.

MAIN FULCRUM

TO BRAKE CYL.

LEADING   TRAILING

TO BRAKE CYL.

BRAKE ARRANGEMENT as for.....

Jubilee 4-4-0 compound
John Hick 2-2-2-2
Greater Britain 2-2-2-2
Jeanie Deans 2-2-2-0
Alfred the Great 4-4-0 compds.
Dreadnought 2-2-2-0 compds.

Brake Arrangements

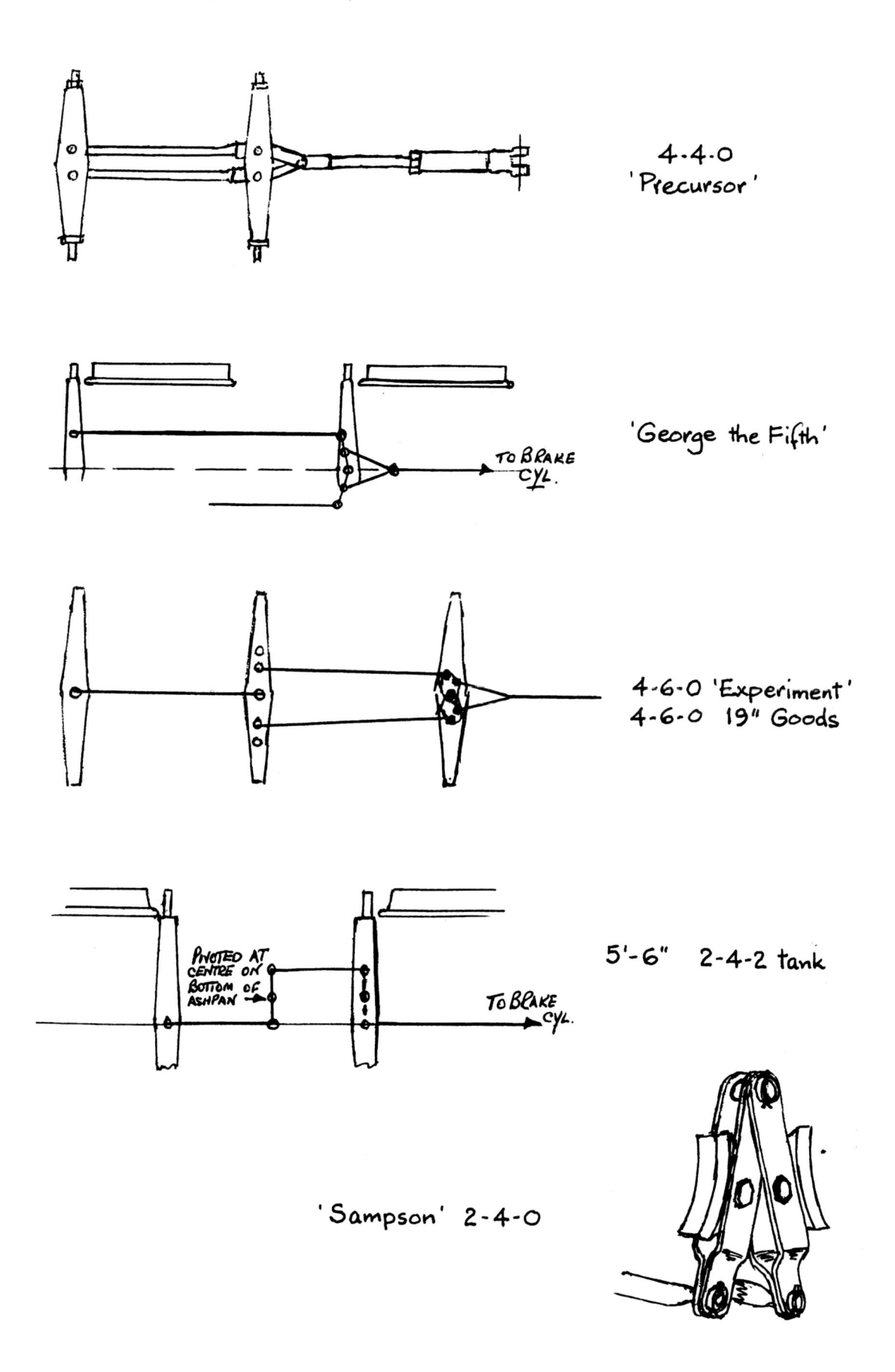

4-4-0 'Precursor'

'George the Fifth'

TO BRAKE CYL.

4-6-0 'Experiment'
4-6-0 19" Goods

PIVOTED AT CENTRE ON BOTTOM OF ASHPAN

TO BRAKE CYL.

5'-6" 2-4-2 tank

'Sampson' 2-4-0

1800 Gallon TENDERS
Build up of tender plates

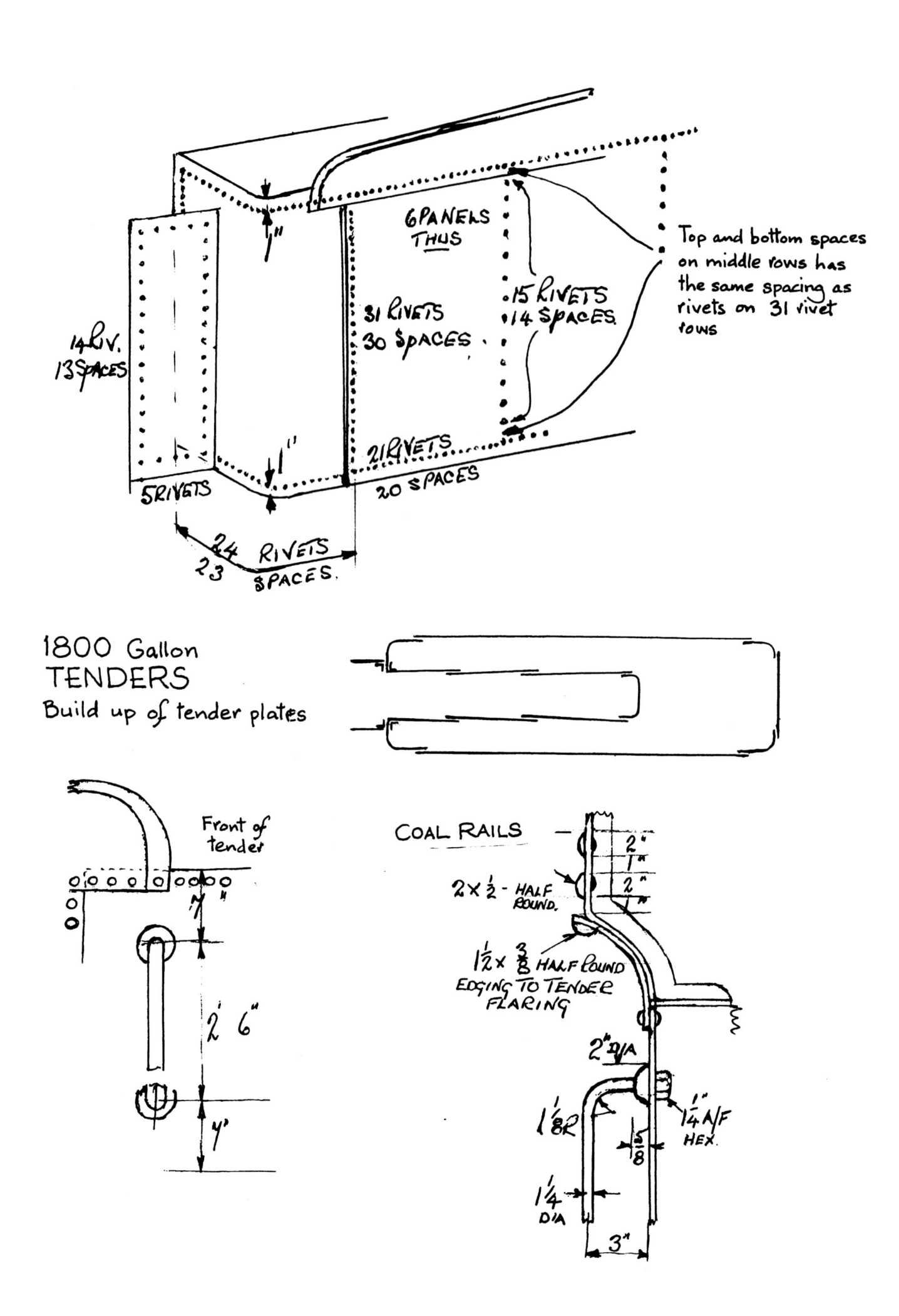

COAL RAILS

1800 Gallon Tenders

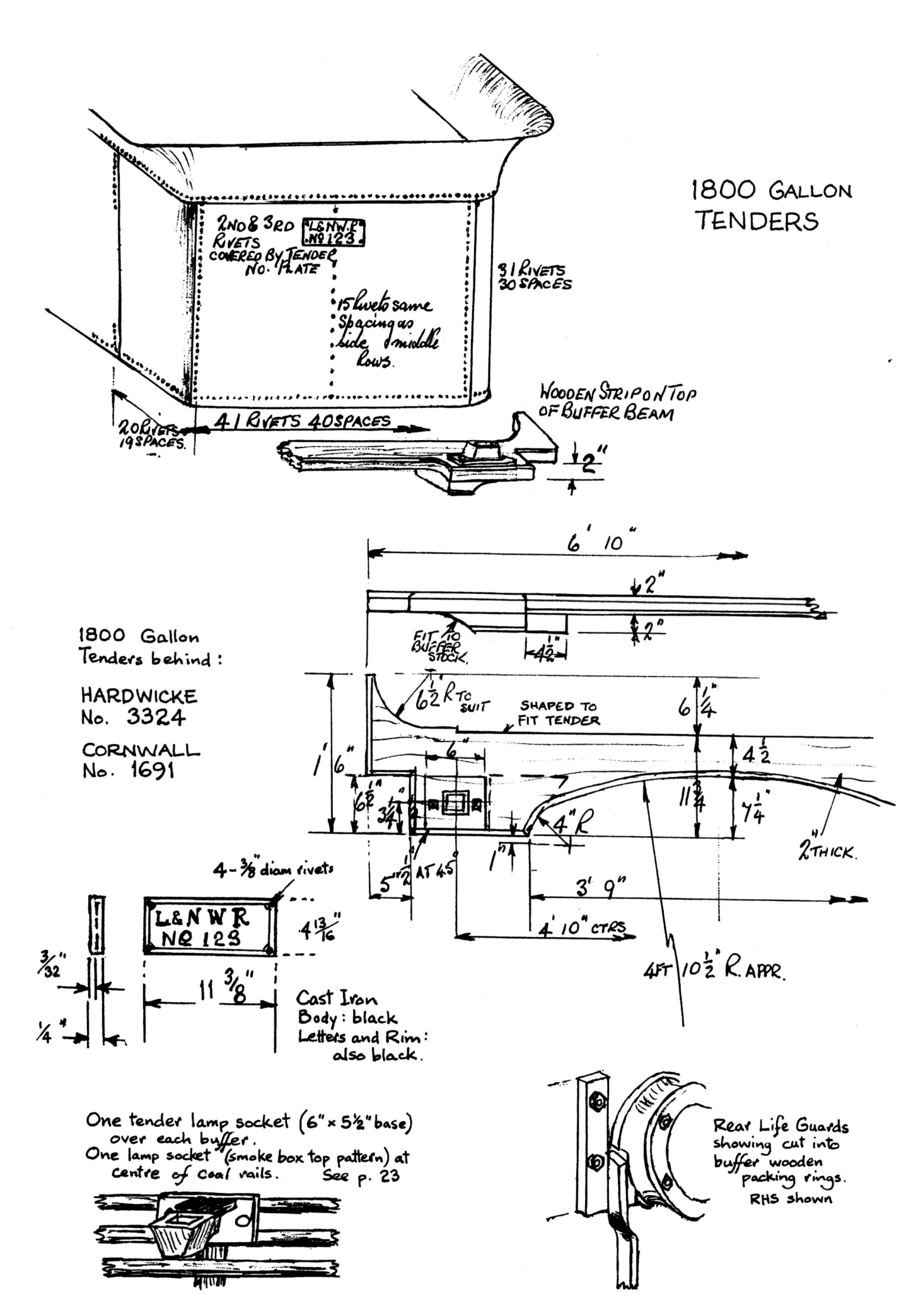

2ND & 3RD RIVETS COVERED BY TENDER No. PLATE

L&NWR No 123

31 RIVETS 30 SPACES

15 Rivets same spacing as side & middle rows.

20 RIVETS 19 SPACES.

41 RIVETS 40 SPACES

WOODEN STRIP ON TOP OF BUFFER BEAM — 2"

1800 Gallon Tenders behind:

HARDWICKE No. 3324

CORNWALL No. 1691

6' 10"
2"
2"
FIT TO BUFFER STOCK. 4½"
6½" R TO SUIT
SHAPED TO FIT TENDER
6¼"
4½"
1' 6"
6½"
¾"
6"
4" R
11¾"
7¼"
2" THICK
5½" AT 45°
1"
3' 9"
4' 10" CTRS
4FT 10½" R. APPR.

4 – ⅜" diam rivets

L&NWR No 123

4 13/16"
11 3/8"
3/32"
¼"

Cast Iron
Body: black
Letters and Rim: also black.

One tender lamp socket (6" × 5½" base) over each buffer.
One lamp socket (smoke box top pattern) at centre of coal rails. See p. 23

Rear Life Guards showing cut into buffer wooden packing rings. RHS shown

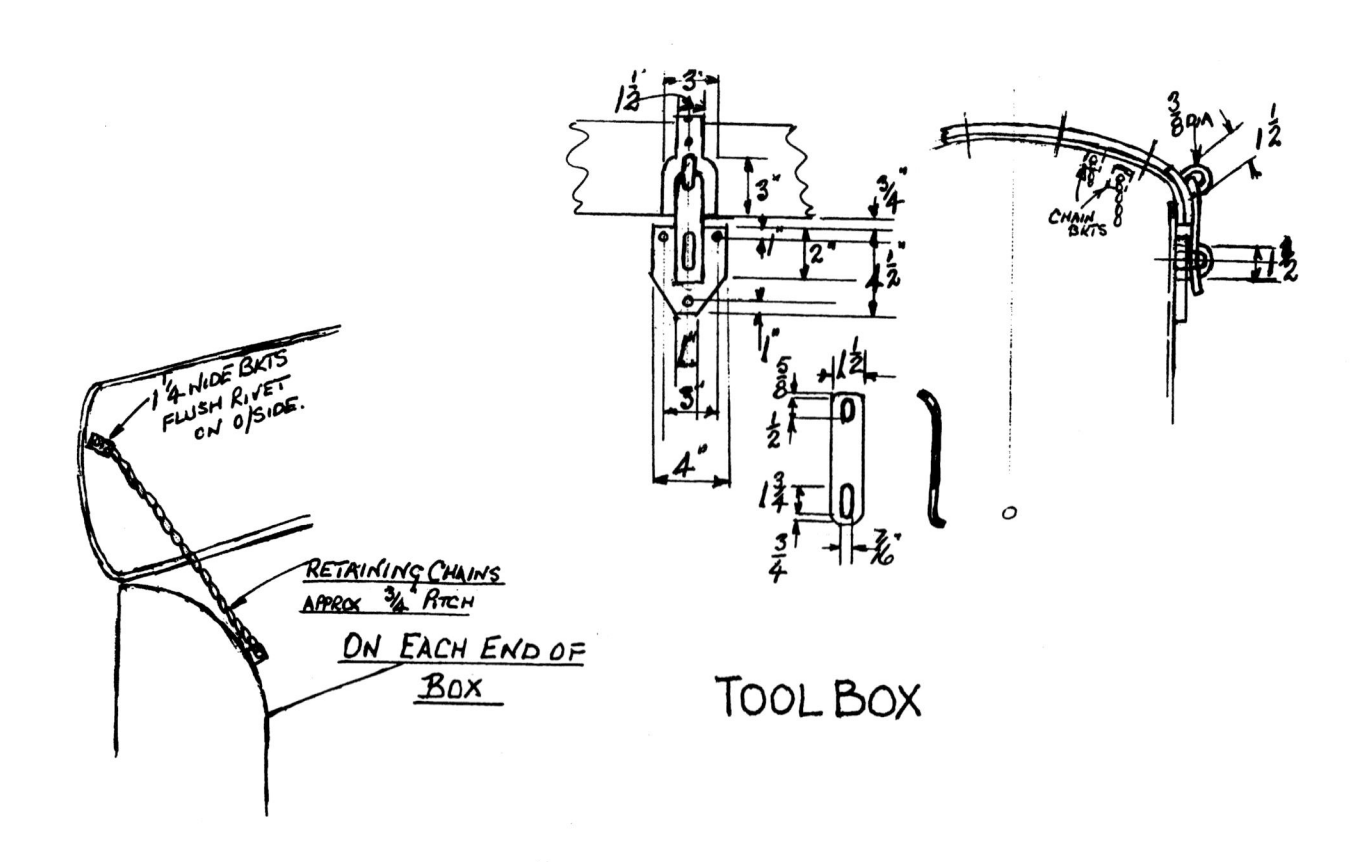

TOOL BOX

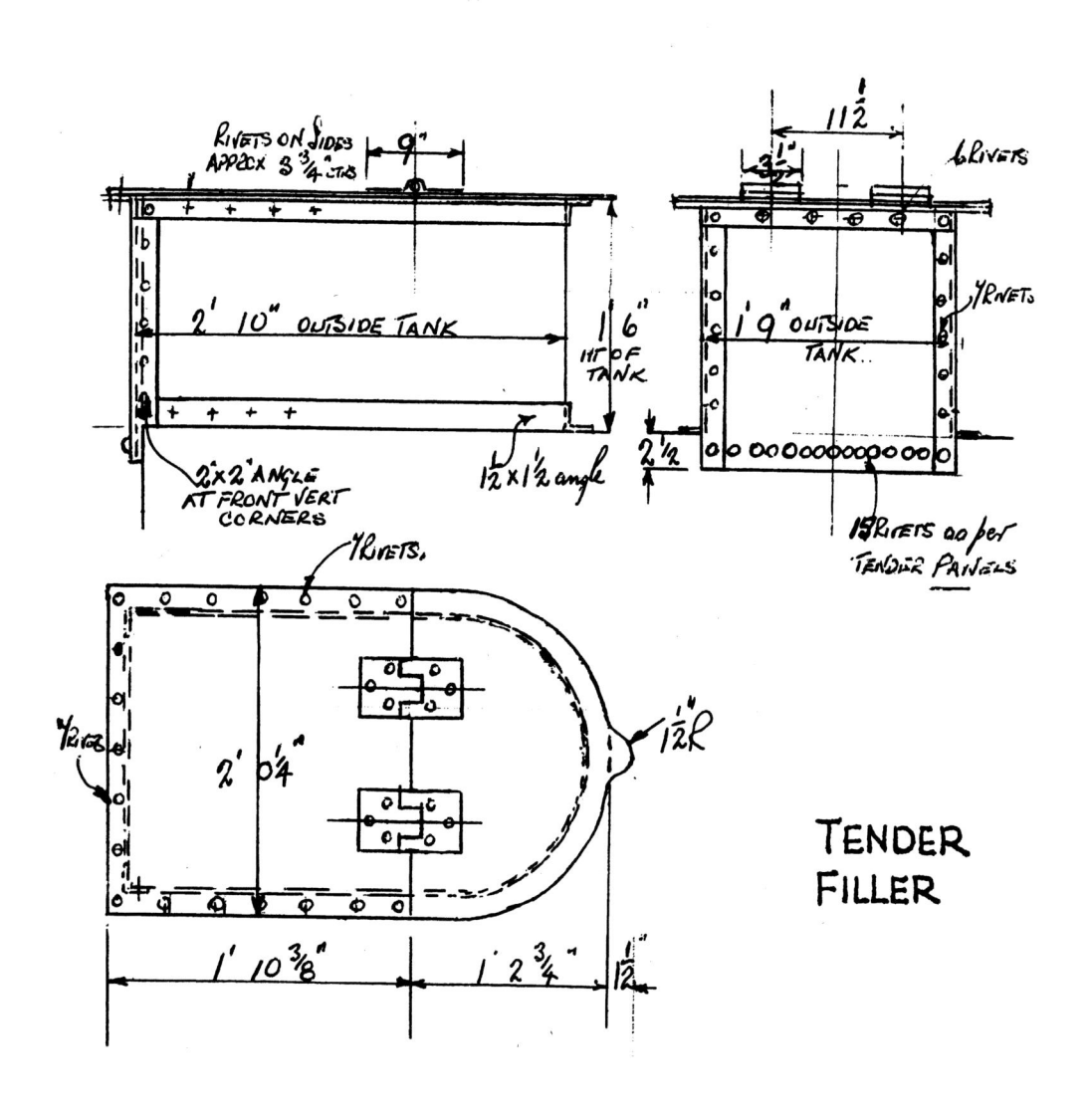

1¼" WIDE BKTS FLUSH RIVET ON O/SIDE.

RETAINING CHAINS APPROX ¾" PITCH

ON EACH END OF BOX

TENDER FILLER

RIVETS ON SIDES APPROX 3¾" CTRS

2' 10" OUTSIDE TANK

1' 6" HT OF TANK

2"x2" ANGLE AT FRONT VERT CORNERS

½" x 1½" angle

1' 9" OUTSIDE TANK

6 RIVETS

4 RIVETS

15 RIVETS ⌀⌀ per TENDER PANELS

2½"

2' 0¼"

1' 10⅜"   1' 2¾"   ½"

1½"R

TOOL BOX

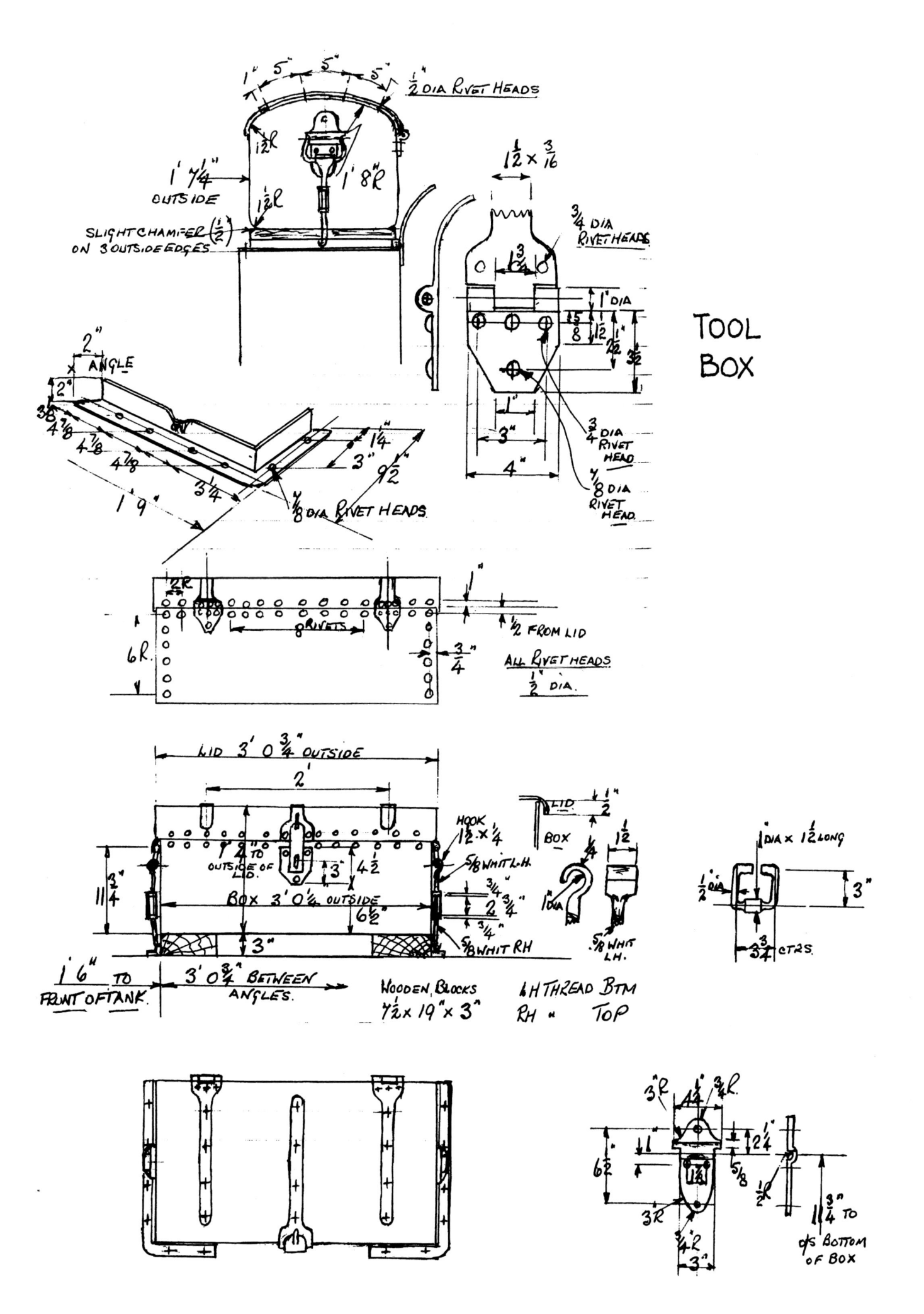

Adjustment of Tender Brake Gear

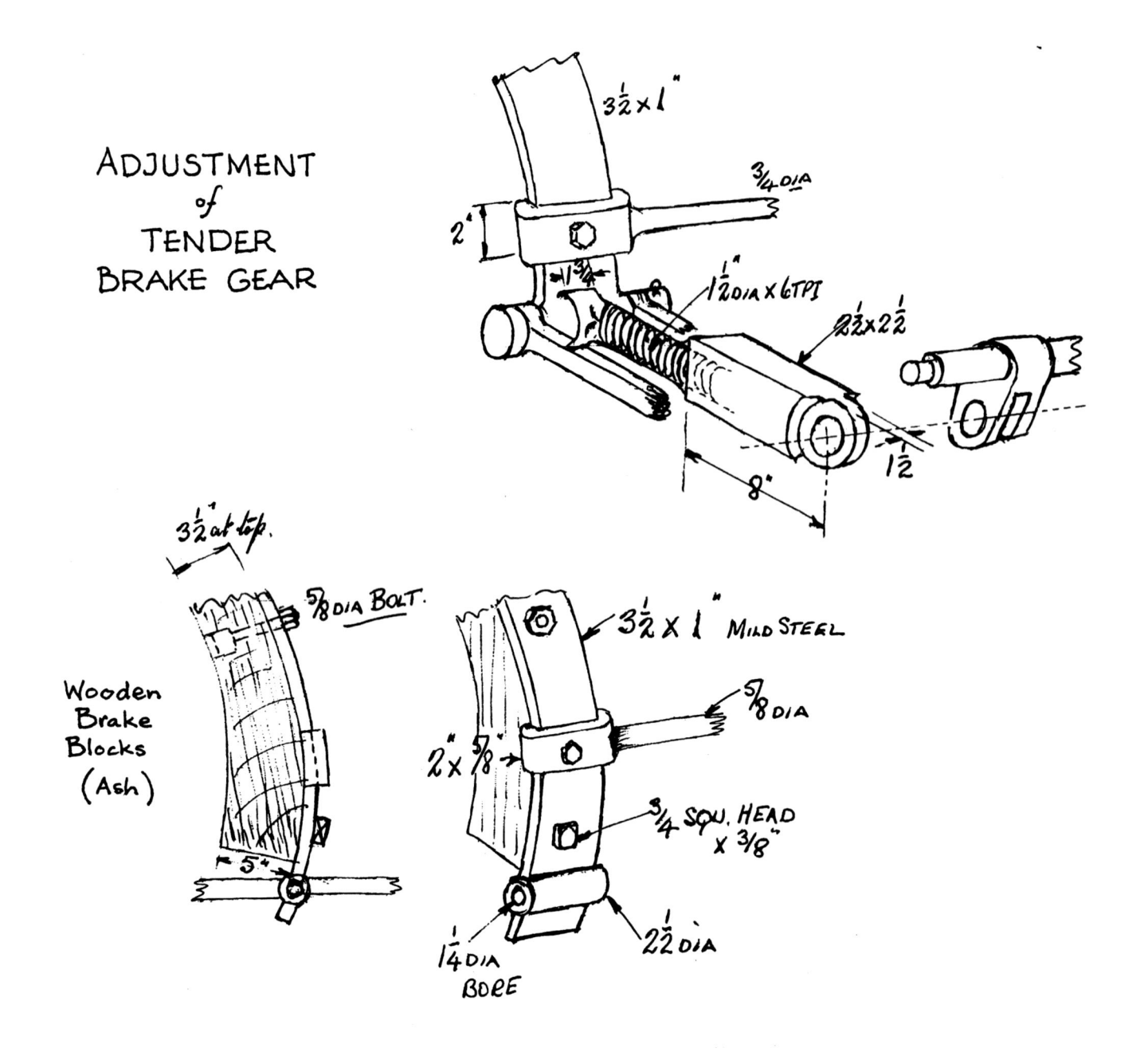

$3\frac{1}{2} \times 1"$
$\frac{3}{4}$ DIA
2"
$1\frac{3}{4}"$
$1\frac{1}{2}"$ DIA $\times$ 6 TPI
$2\frac{1}{2} \times 2\frac{1}{2}$
$1\frac{1}{2}"$
8"

Wooden Brake Blocks (Ash)

$3\frac{1}{2}"$ at top.
$\frac{5}{8}$ DIA BOLT.
5"

$3\frac{1}{2} \times 1"$ MILD STEEL
$\frac{5}{8}$ DIA
$2 \times \frac{5}{8}"$
$\frac{3}{4}$ SQU. HEAD $\times \frac{3}{8}"$
$1\frac{1}{4}$ DIA BORE
$2\frac{1}{2}$ DIA

Pipe brackets behind tender steps.
RHS. drawn

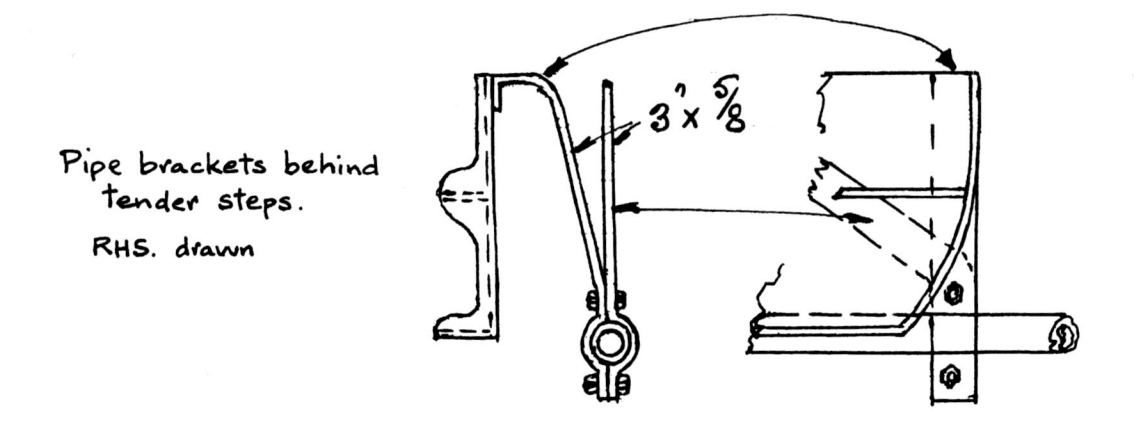

$3" \times \frac{5}{8}"$

TENDER WATER LEVEL INDICATOR

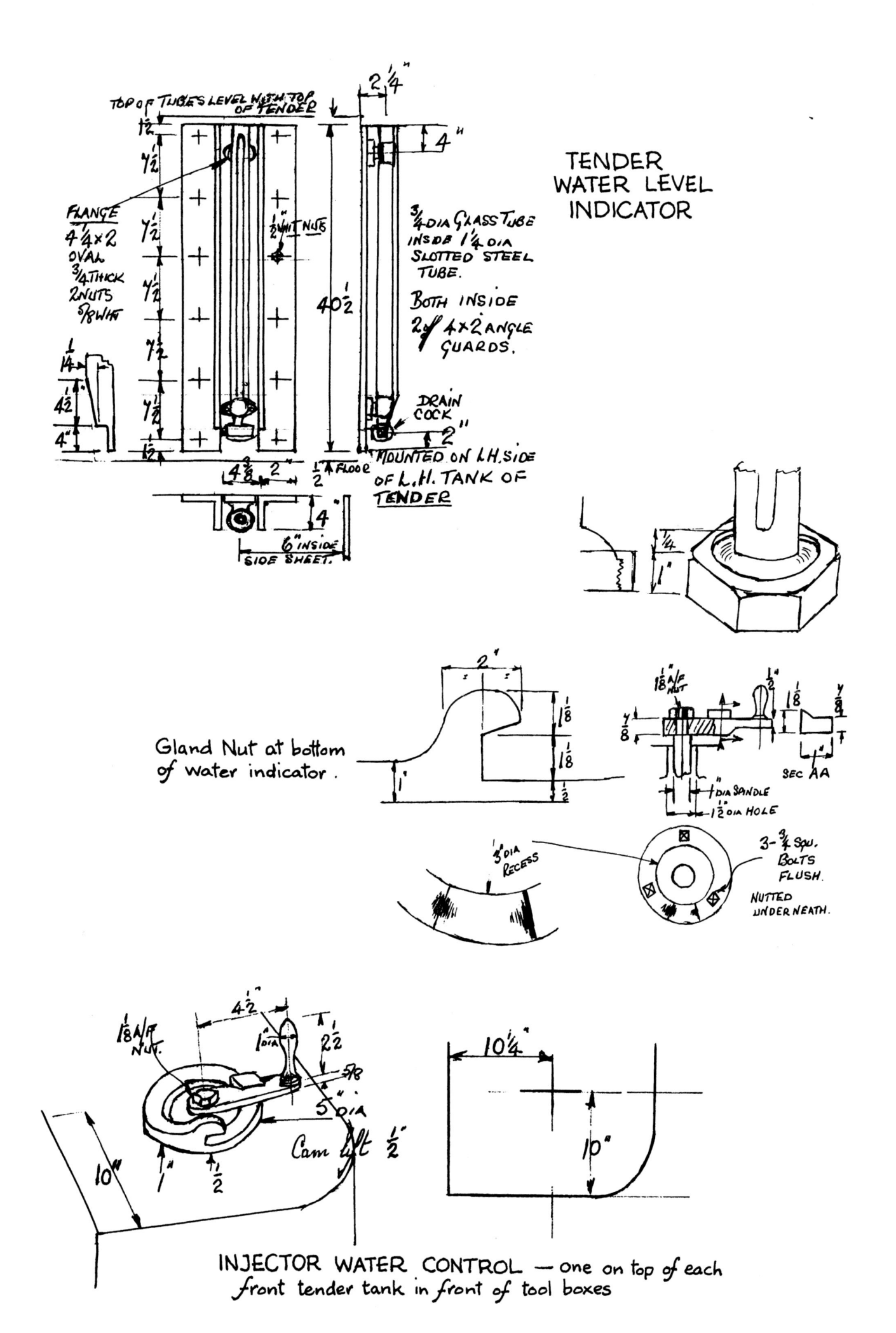

Gland Nut at bottom of water indicator.

INJECTOR WATER CONTROL — one on top of each front tender tank in front of tool boxes

TENDER PICK-UP HANDLE and FOOT SUPPORT TREADLE

TENDER HAND BRAKE CASTING

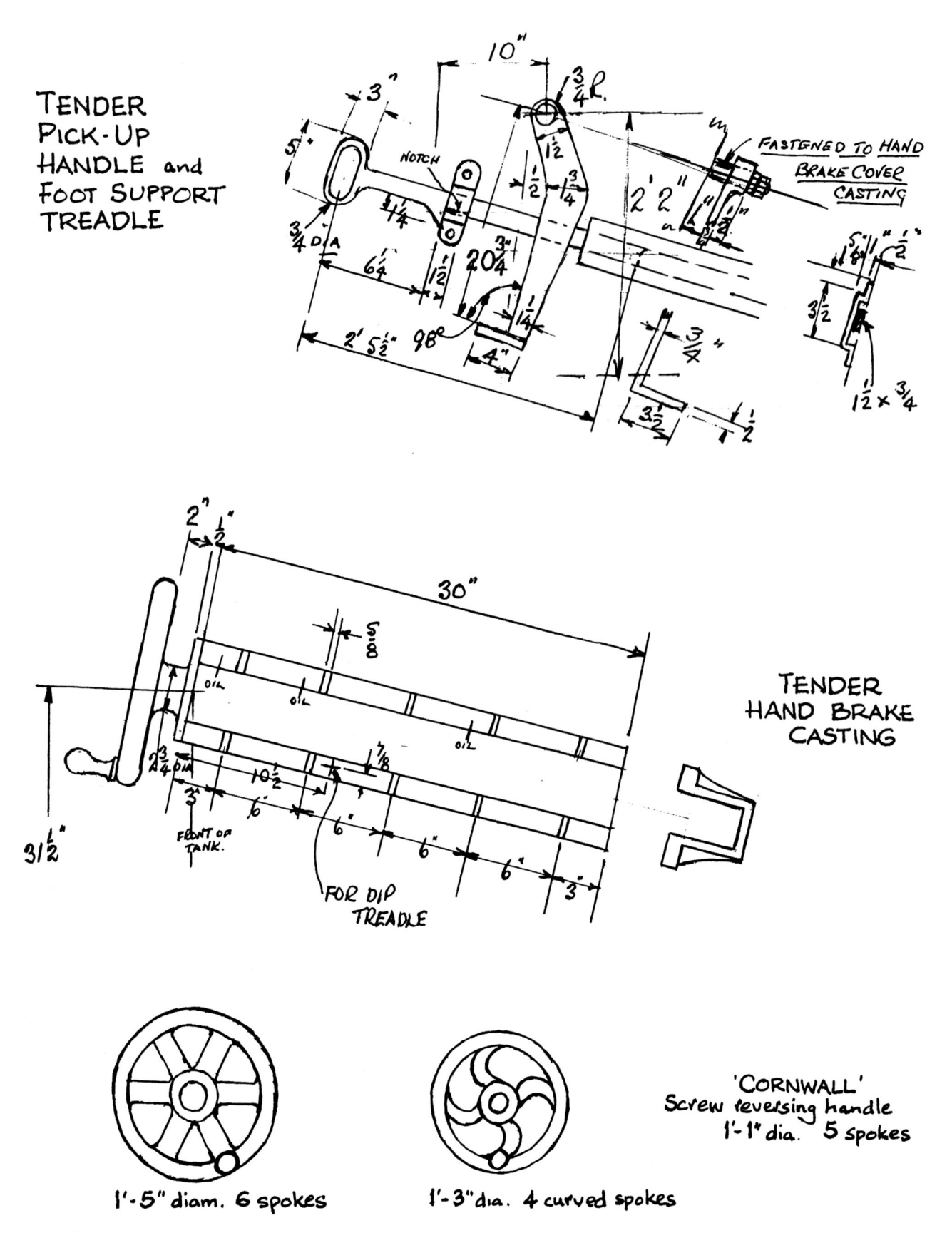

'CORNWALL'
Screw reversing handle
1'-1" dia. 5 spokes

1'-5" diam. 6 spokes

1'-3" dia. 4 curved spokes

ALTERNATIVE TYPES OF SCREW REVERSE AND TENDER HAND BRAKE WHEELS AS FITTED TO EARLIER TYPES OF ENGINES, INCLUDING 'PRECEDENTS' IN THEIR EARLY LIFE.

WATER SCOOP
1800 gal. tender

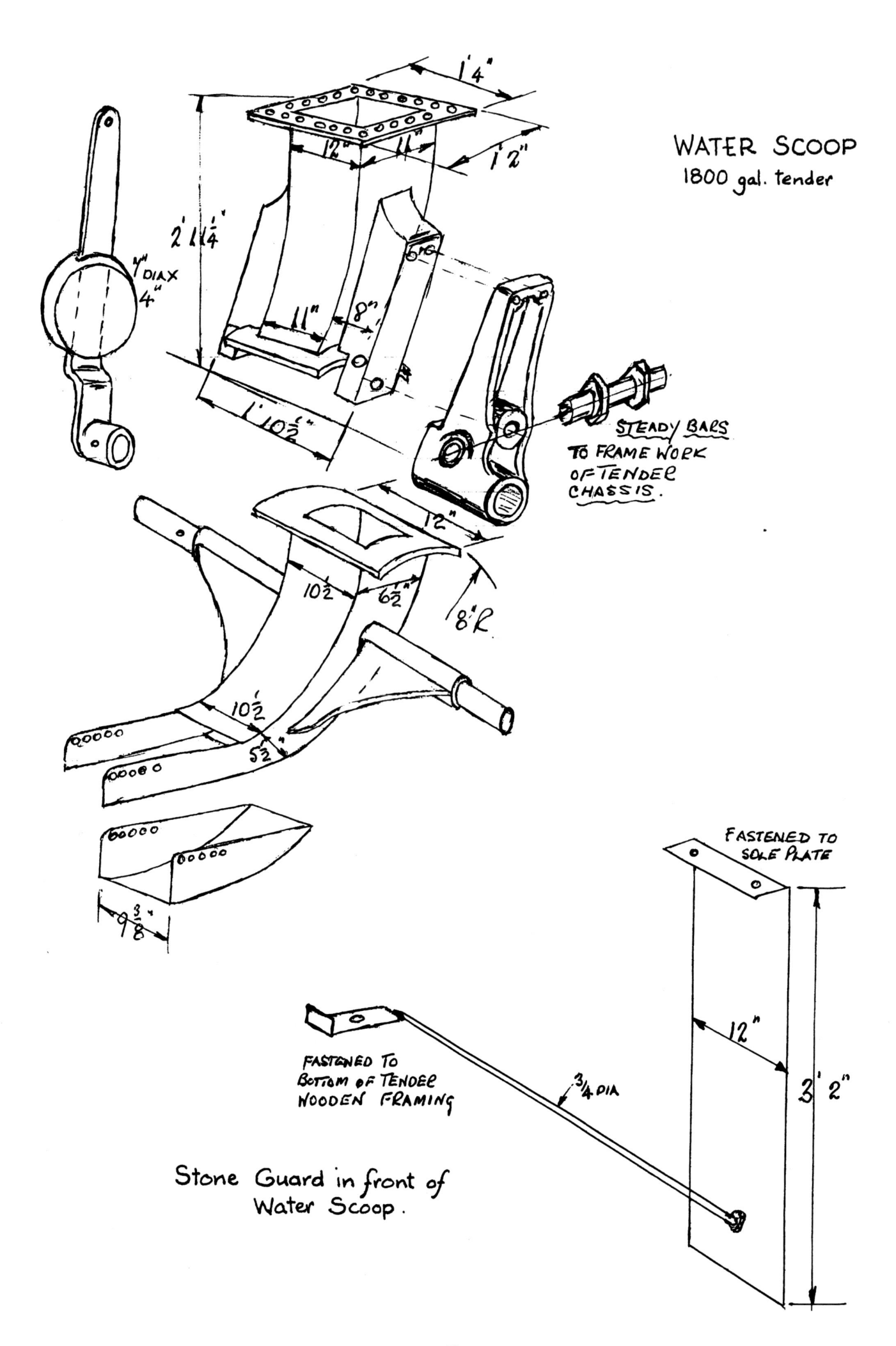

Steady bars to framework of tender chassis.

Stone Guard in front of Water Scoop.

Fastened to sole plate

Fastened to bottom of tender wooden framing

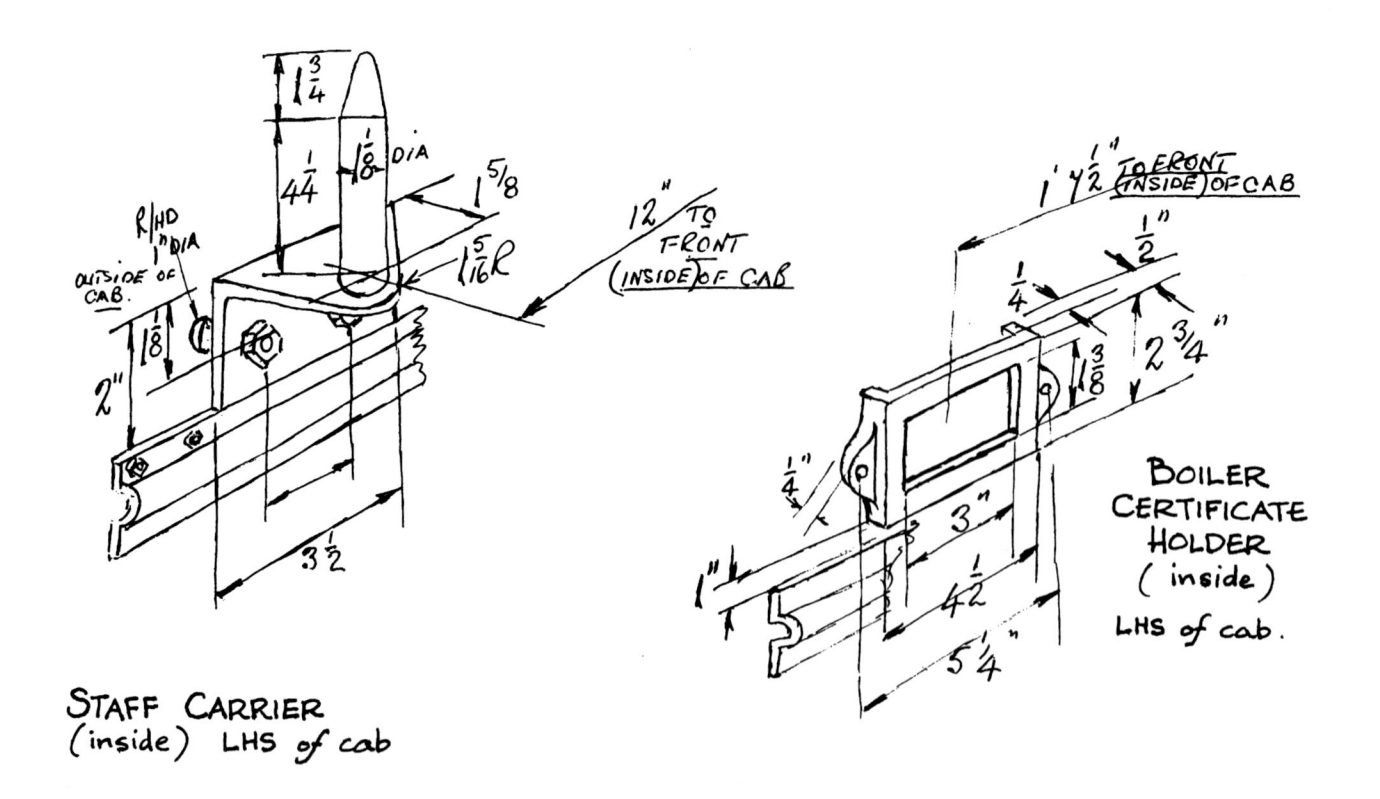

STAFF CARRIER
(inside) LHS of cab

BOILER CERTIFICATE HOLDER
(inside) LHS of cab.

HOOK on inside of tender LHS.
to hold flap plate up.

GAUGE GLASS LAMP BRACKET
on tray shelf

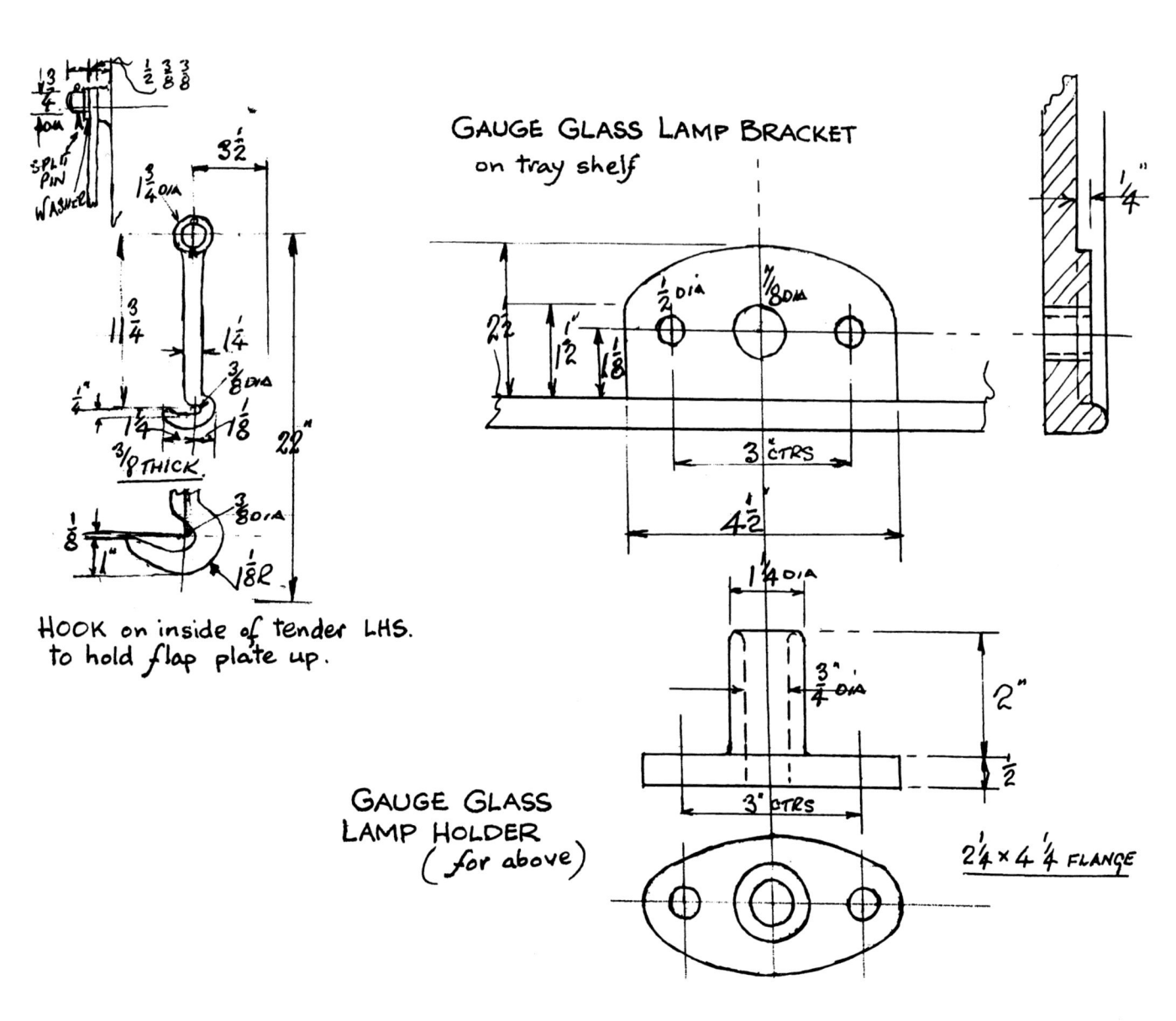

GAUGE GLASS LAMP HOLDER
(for above)

$2\tfrac{1}{4} \times 4\tfrac{1}{4}$ FLANGE

Gauge Glass Lamp

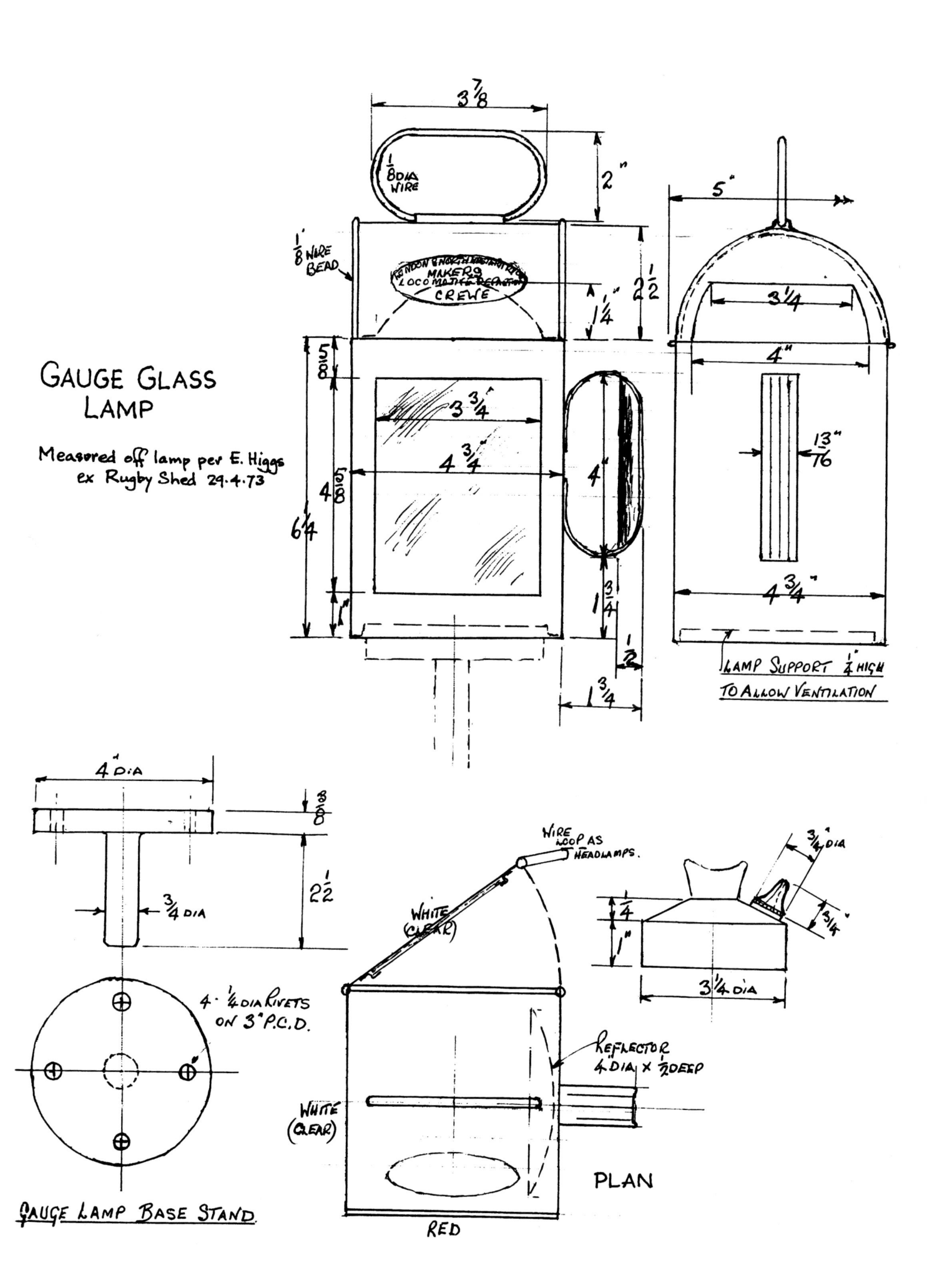

Measured off lamp per E. Higgs
ex Rugby Shed 29.4.73

Gauge Lamp Base Stand

Plan

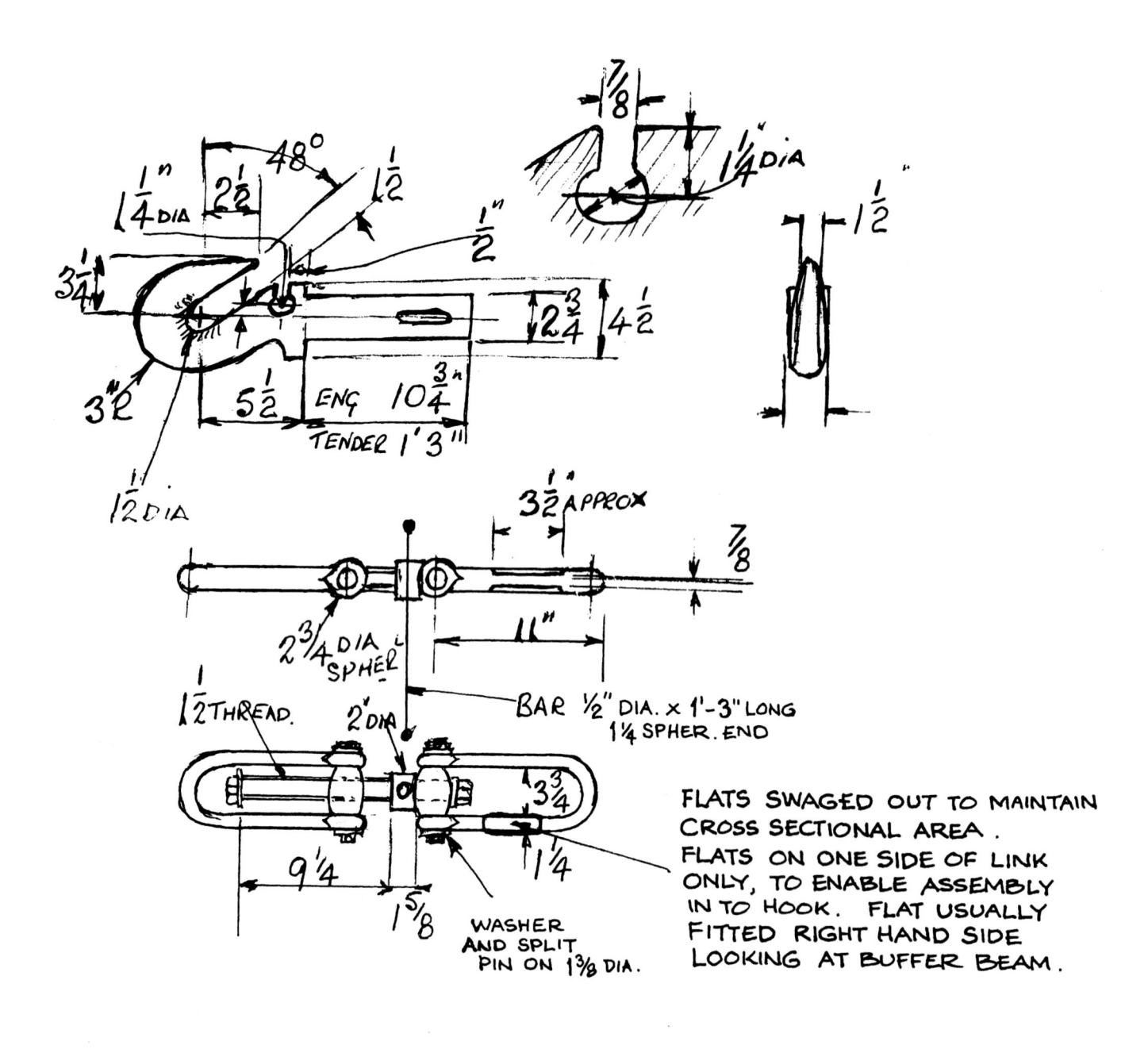

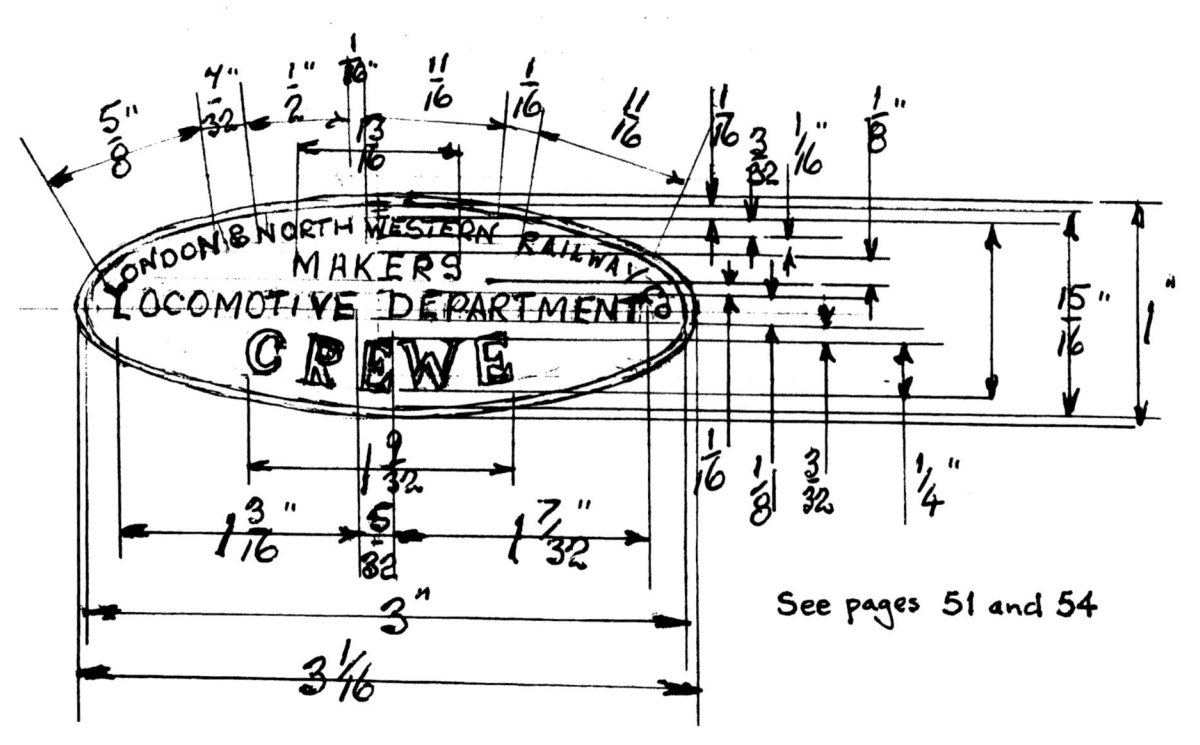

FLATS SWAGED OUT TO MAINTAIN CROSS SECTIONAL AREA.
FLATS ON ONE SIDE OF LINK ONLY, TO ENABLE ASSEMBLY IN TO HOOK. FLAT USUALLY FITTED RIGHT HAND SIDE LOOKING AT BUFFER BEAM.

See pages 51 and 54

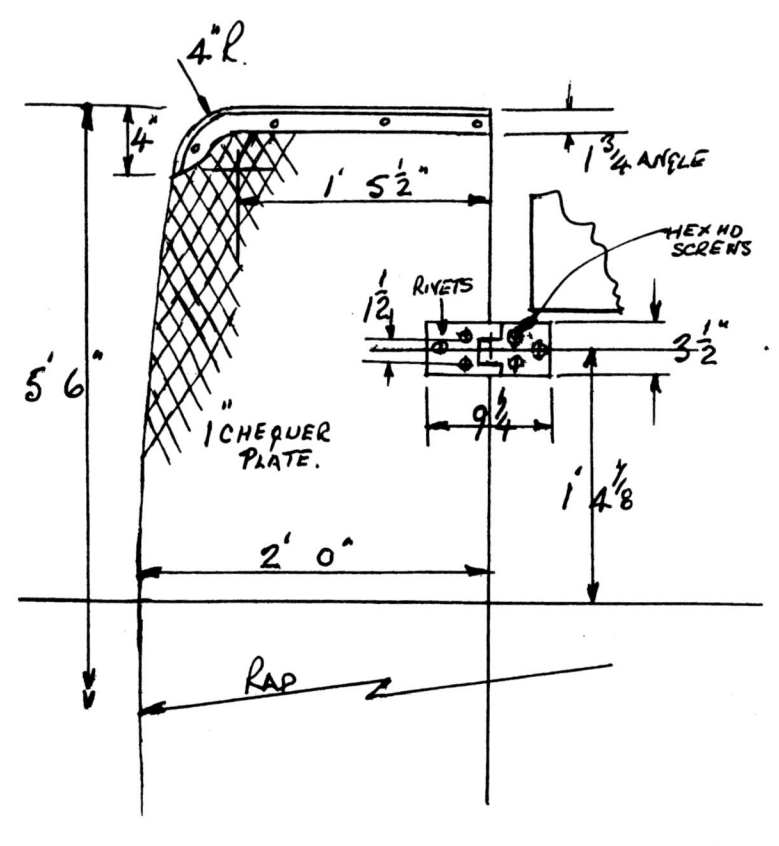

TENDER FLAP PLATE

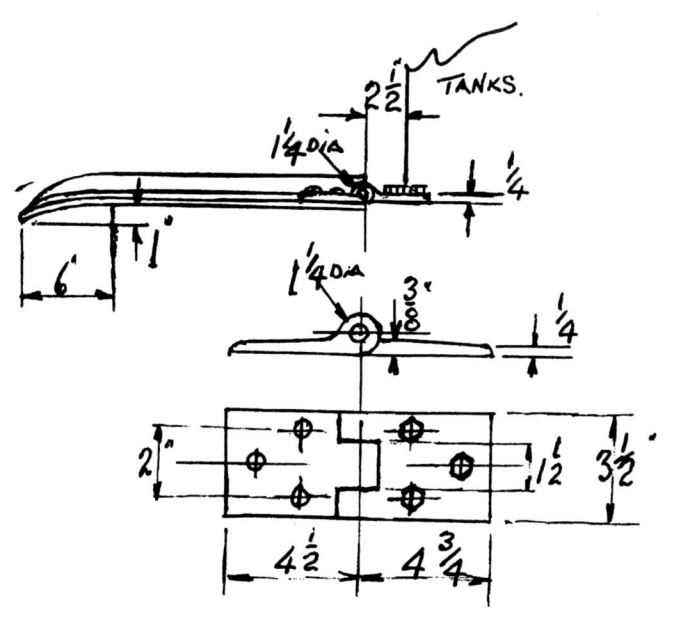

TRAIN HEATING PIPE
ON RIGHT HAND SPLASHER IN CAB

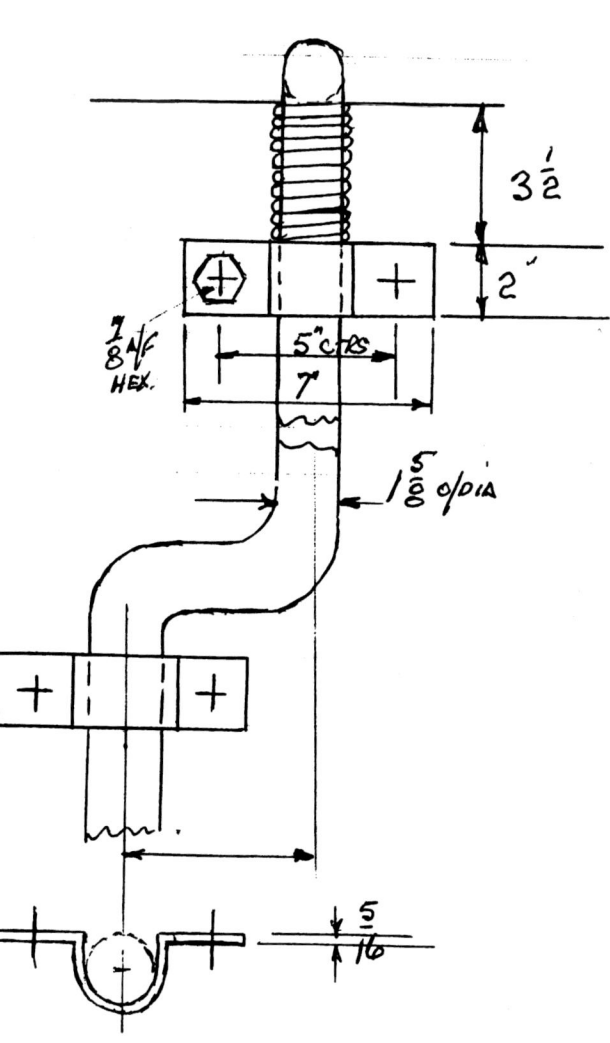

Hardwick Lining

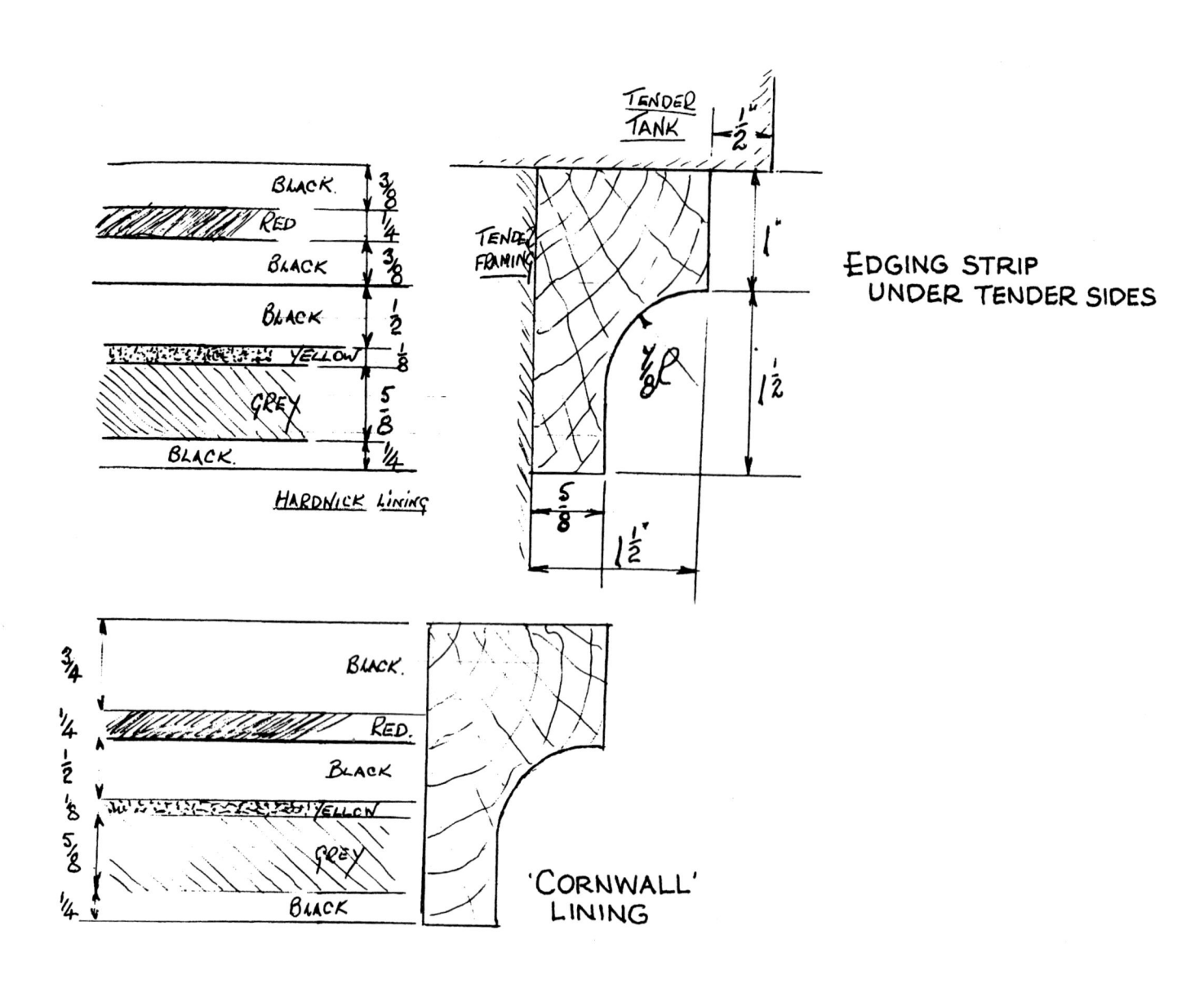

- Black — 3/8
- Red — 1/4
- Black — 3/8
- Black — 1/2
- Yellow — 1/8
- Grey — 5/8
- Black — 1/4

Edging Strip Under Tender Sides

Tender Tank — 1/2"
Tender Framing
1"
7/8 R
1 1/2"
5/8
1 1/2"

'Cornwall' Lining

- 3/4 Black
- 1/4 Red
- 1/2 Black
- 1/8 Yellow
- 5/8 Grey
- 1/4 Black

Ramsbottom Locomotive Headlamp (Hambleton)

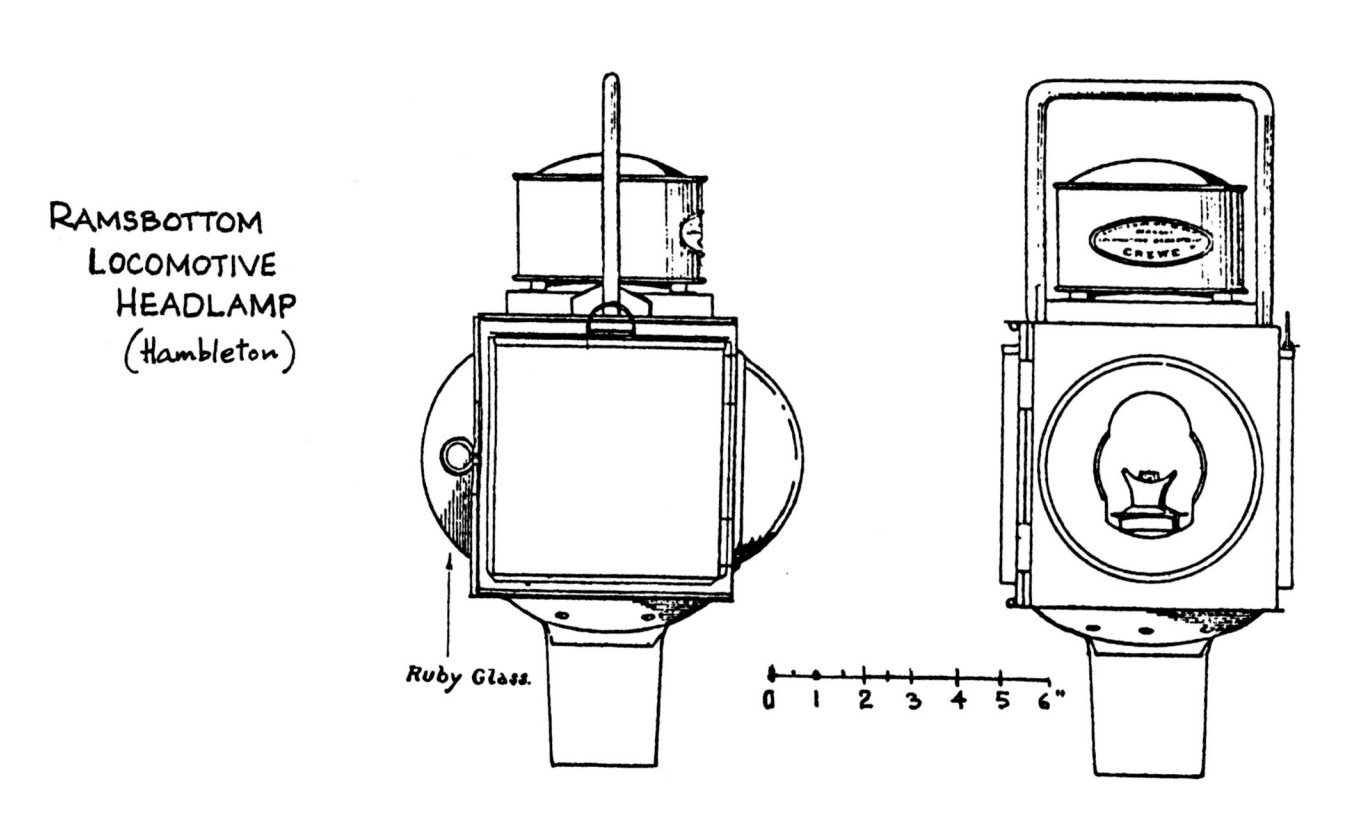

Ruby Glass.

0 1 2 3 4 5 6"

When looking at photographs of Bill Finch's model like this, one has to remember that, though only 5-inch gauge, the model is live steam and the controls actually work. The attention to detail is quite staggering. The horizontal bar coupling the vacuum brake control on the left with the steam brake control on the right is just above the top of the firebox. (See Pages 9 and 16)
National Railway Museum, York

Here as a direct comparison is the cab of *Hardwicke* as now displayed in the NRM. The cab sides below the waist are red, as painted by the LMS in 1946 and repainted by the NRM in 1976. In Bill Finch's model above these parts are black, the correct LNWR livery.
Simon Fountain

Two more views of Bill Finch's masterpiece, showing the smokebox, valve chest and cylinder fronts, above, and front of the footplate and part of the motion work, left.
Simon Fountain

Three detail views of Hardwicke.

Photos by Ted Talbot

Just in front of the firebox can be seen, right to left, the left-hand driving axlebox, left-hand big end, the eccentric straps and rods and the right-hand crank web. Top right is the boiler mud hole. Just behind it and top left are two of the boiler band clamps. As with the lower cab on Page 55, the frame components painted red in this view would have been black in original LNWR condition.

The ash hopper, the cylinder drain cocks and their operating levers.

The loco rear steps and a tender front buffer.

Index

Ashpan damper handle	14
Blower handles	15
Boiler bands	33
Boiler certificate holder	50
Brake adjustment (tender)	41
Brake arrangements	40
Brake compensation	40
Buffers	34
Cab handrails	20
Cab roof	13
Cab side arrangement	21
Cab spectacles	19
Carriage warming valve	22
Chimneys	23
Coal rails	42
Couplings	52
Compensation of brakes	40
Connecting rod	35
Crosshead	35
Dewrance water gauge	17
Domes	24
Drain cock control	18
Edging strip (tender)	54
Engine steps	21
Fire glare shield	14
Fire hole door	17
Fly sheet hooks	13
Footplate	9
Front buffer beam lamp sockets	28
Front springs	31
Front vacuum pipe	29
Gauge glass lamp	18
Gauge glass lamp stand	51
Guide bars	36
Hand rails	20
Hand rail knobs	39
Headlamp	54
Holder for boiler report	13
Injector water control	47
Lamp socket (buffer beam)	28
Lamp socket (chimney)	27
Lamp socket (tender)	27
Leg shield	14
Lubricators	26/27
Method of holding cabside to bottom	20
Motion	37/38
Motion plate	36
Number plate	18
Oil boxes	39
Piston	38
Pressure gauges	15
Rear vacuum pipe	28
Reversing gear	48
Safety hooks	33
Safety valve	25
Sand valve	32
Sandbox covers	32
Screw coupling	52
Screw reverse	48
Shed plate	12
Sight feed lubricator	19
Smokebox door dogs	31
Smoke box door fastening	30
Smokebox door hinge	30
Smokebox regulator lubricator	26
Snifting valves	26
Springs	31
Square splasher top	20
Staff carrier	50
Steam chest lubricator	14
Stone guard	49
Tender controls	10
Tender filler	44
Tender fittings	47
Tender number plate	43
Toolbox	44
Train heating pipe	53
Vacuum brake control	16
Vacuum ejector pipe support	39
Vacuum pipe bracket	29
Valve motion	38
Water scoop	49
Wheel spokes	33
Whistle	22
Whistle handle	9
Window catches	13
Wooden brake blocks	46